Macmillan/McGraw-Hill • Glencoe

Grade 3

Math Triumphs

Book 2: Number and Operations

Authors

Basich Whitney • Brown • Dawson • Gonsalves • Silbey • Vielhaber

Mc Graw Hill Macmillan/McGraw-Hill
Glencoe

Photo Credits

cov Alloy Photography/Veer; **iv** (tl) File Photo, (tc) The McGraw-Hill Companies, (tr) The McGraw-Hill Companies, (cl) Doug Martin, (c) Doug Martin, (cr) Doug Martin, (bl) File Photo, (bc) File Photo; **vi** Richard Hutchings/Digital Light Source; **vii** Richard Hutchings/Digital Light Source; **viii** Richard Hutchings/Digital Light Source; **119** Richard Hutchings/Digital Light Source; **121** (tl) The McGraw-Hill Companies Inc./Ken Cavanagh Photographer, (tr) The McGraw-Hill Companies, (b) Russell Monk/Masterfile; **125** Ed-Imaging; **127** Ed-Imaging; **131** Ed-Imaging; **134** The McGraw-Hill Companies, Inc./ Jack Holtel Photographer; **135** The McGraw-Hill Companies Inc./Ken Cavanagh Photographer; **137** Lew Robertson/CORBIS; **140** Richard Hutchings/Digital Light Source; **144** Richard Hutchings/Digital Light Source.; **147** Richard Hutchings/Digital Light Source; **148** (tl) Comstock Images/Alamy Images, (tc) Siede Preis/Getty Images, (tr) Getty Images, (b) Burke/Triolo Productions/Getty Images; **151** Westend61/Alamy Images; **153** Richard Hutchings/Digital Light Source; **159** D. Hurst/Alamy Images; **165** (tl) Getty Images, (tcl) Stockdisc/PunchStock, (tcr) Stockdisc/PunchStock, (tr) Getty Images; **166** (tl) Jeremy Woodhouse/Getty Images, (tcl) Tom Brakefield/Getty Images; **170** (t) Jupiterimages/Alamy Images, (b) 2006 Photos To Go; **173** Masterfile; **175** 2006 Photos To Go; **182** Masterfile.; **185** Richard Hutchings/Digital Light Source; **187** (t) Joe Atlas/Jupiterimages, (b) D. Hurst/Alamy Images; **188** Thomas Northcut/Getty Images; **189** Emanuele Taroni/Getty Images; **191** Richard Hutchings/Digital Light Source; **193** Ingram Publishing/Fotosearch; **197** Blend Images/Alamy Images; **199** CORBIS; **200** C Squared Studios/Getty Images; **201** The McGraw-Hill Companies Inc./Ken Cavanagh Photographer; **202** (t) The McGraw-Hill Companies, (b) G.K. Vikki Hart/Getty Images; **203** (tl) Getty Images, (tr) Stockdisc/PunchStock, (b) D. Hurst/Alamy Images; **206** Siede Preis/Getty Images; **207** Brand X/CORBIS; **209** United States coin images from the United States Mint; **210** (tl) Stockdisc/PunchStock, (tc) Stockdisc/PunchStock, (b) Tony Freeman/PhotoEdit; **211** Eclipse Studios; **213** (l) Index Stock/Alamy Images, (r) Stock Food/SuperStock; **214** Masterfile; **218** Getty Images; **220** Richard Hutchings/Digital Light Source.

The McGraw·Hill Companies

 Macmillan/McGraw-Hill
Glencoe

Send all inquiries to:
Macmillan/McGraw-Hill • Glencoe/McGraw-Hill
8787 Orion Place
Columbus, OH 43240-4027

ISBN: 978-0-07-888199-2
MHID: 0-07-888199-4

Printed in the United States of America.

3 4 5 6 7 8 9 10 066 16 15 14 13 12 11 10 09 08

Math Triumphs
Grade 3, Book 2

Math Triumphs

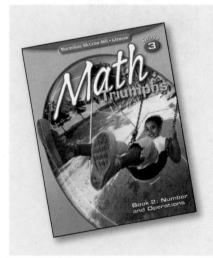
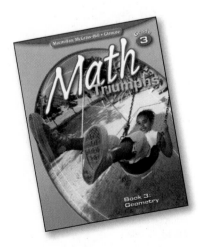

Authors and Consultants

CONSULTING AUTHORS

Frances Basich Whitney
Project Director, Mathematics K–12
Santa Cruz County Office of Education
Capitola, California

Kathleen M. Brown
Math Curriculum Staff Developer
Washington Middle School
Long Beach, California

Dixie Dawson
Math Curriculum Leader
Long Beach Unified
Long Beach, California

Philip Gonsalves
Mathematics Coordinator
Alameda County Office of Education
Hayward, California

Robyn Silbey
Math Specialist
Montgomery County Public Schools
Gaithersburg, Maryland

Kathy Vielhaber
Mathematics Consultant
St. Louis, Missouri

CONTRIBUTING AUTHORS

Viken Hovsepian
Professor of Mathematics
Rio Hondo College
Whittier, California

FOLDABLES
Study Organizer

Dinah Zike
Educational Consultant
Dinah-Might Activities, Inc.
San Antonio, Texas

CONSULTANTS

Assessment

Donna M. Kopenski, Ed.D.
Math Coordinator K–5
City Heights Educational Collaborative
San Diego, California

Instructional Planning and Support

Beatrice Luchin
Mathematics Consultant
League City, Texas

ELL Support and Vocabulary

ReLeah Cossett Lent
Author/Educational Consultant
Alford, Florida

Reviewers

Each person reviewed at least two chapters of the Student Study Guide, providing feedback and suggestions for improving the effectiveness of the mathematics instruction.

Dana M. Addis
Teacher Leader
Dearborn Public Schools
Dearborn, MI

Renee M. Blanchard
Elementary Math Facilitator
Erie School District
Erie, PA

Jeanette Collins Cantrell
5th and 6th Grade Math Teacher
W.R. Castle Memorial Elementary
Wittensville, KY

Helen L. Cheek
K-5 Math Specialist
Durham Public Schools
Durham, NC

Mercy Cosper
1st Grade Teacher
Pershing Park Elementary
Killeen, TX

Bonnie H. Ennis
Math Coordinator
Wicomico County Public Schools
Salisbury, MD

Sheila A. Evans
Instructional Support Teacher - Math
Glenmount Elementary/Middle School
Baltimore, MD

Lisa B. Golub
Curriculum Resource Teacher
Millennia Elementary
Orlando, FL

Donna Hagan
Program Specialist - Special Programs
    Department
Weatherford ISD
Weatherford, TX

Russell Hinson
Teacher
Belleview Elementary
Rock Hill, SC

Tania Shepherd Holbrook
Teacher
Central Elementary School
Paintsville, KY

Stephanie J. Howard
3rd Grade Teacher
Preston Smith Elementary
Lubbock, TX

Rhonda T. Inskeep
Math Support Teacher
Stevens Forest Elementary School
Columbia, MD

Albert Gregory Knights
Teacher/4th Grade/Math Lead Teacher
Cornelius Elementary
Houston, TX

Barbara Langley
Math/Science Coach
Poinciana Elementary School
Kissimmee, FL

David Ennis McBroom
Math/Science Facilitator
John Motley Morehead Elementary
Charlotte, NC

Jan Mercer, MA; NBCT
K-5 Math Lab Facilitator
Meadow Woods Elementary
Orlando, FL

Rosalind R. Mohamed
Instructional Support Teacher - Math
Furley Elementary School
Baltimore, MD

Patricia Penafiel
Teacher
Phyllis Miller Elementary
Miami, FL

Lindsey R. Petlak
2nd Grade Instructor
Prairieview Elementary School
Hainesville, IL

Lana A. Prichard
District Math Resource Teacher K-8
Lawrence Co. School District
Louisa, KY

Stacy L. Riggle
3rd Grade Spanish Magnet Teacher
Phillips Elementary
Pittsburgh, PA

Wendy Scheleur
5th Grade Teacher
Piney Orchard Elementary
Odenton, MD

Stacey L. Shapiro
Teacher
Zilker Elementary
Austin, TX

Kim Wilkerson Smith
4th Grade Teacher
Casey Elementary School
Austin, TX

Wyolonda M. Smith, NBCT
4th Grade Teacher
Pilot Elementary School
Greensboro, NC

Kristen M. Stone
3rd Grade Teacher
Tanglewood Elementary
Lumberton, NC

Jamie M. Williams
Math Specialist
New York Mills Union Free School District
New York Mills, NY

Contents

CHAPTER 4 — Place Value

CHAPTER 5 Fractions

Contents

CHAPTER 6 — Fraction Equivalence

Home Connection

English

Spanish

Dear Family,

Today our class started **Chapter 4, Place Value.** In this chapter, I will learn how to count and estimate numbers less than 100. I will also learn to represent numbers in standard form, word form, and expanded form. I will learn how to model numbers to show thousands, hundreds, tens, and ones.

Love, Nathan

Estimada familia:

Hoy en clase comenzamos el **Capítulo 4, Valor posicional.** En este capítulo aprenderé a contar y a calcular números menores de 100. También, aprenderé a representar números en forma estándar, con palabras y en forma desarrollada. Aprenderé a usar los números para mostrar millares, centenas, decenas y unidades.

Cariños, _____

Help at Home

You can help your child learn about place value by identifying place value of digits. Look for numbers in magazines and newspapers. Ask your child how many thousands, hundreds, tens, or ones are in each number.

Math Online Take the chapter Get Ready quiz at macmillanmh.com.

Ayude en casa

Usted puede ayudar a su hijo(a) a aprender sobre el valor posicional al identificar el valor posicional de los dígitos. Busquen números en revistas y periódicos. Pregunte a su hijo(a) cuántos millares, centenas, decenas y unidades hay en cada número.

Name _____

Get Ready

Write each missing number.

1 1, 2, 3, __4__, 5

2 6, 7, __8__, 9, 10

Write each number shown.

3

__1__

4

__10__

5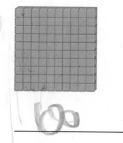

__60__

Write each number.

6 two

_____

7 nine

_____

8 four

_____

Write each number name.

9 3 _____

10 6 _____

Find each sum.

11 20 + 3 = _____

12 80 + 9 = _____

13 Find each number missing from the number line.

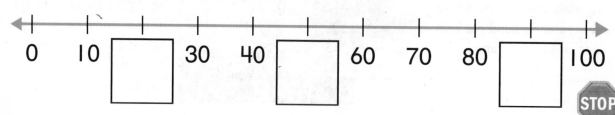

Name _____

Count to 100

Key Concept

There are 35 marbles in all.
There are 3 sets of ten marbles.
There is one set of 5 marbles.

The number 35 has the digit 3 in the **tens** place
and the digit 5 in the **ones** place.

35   **standard form**

tens	ones
3	5

Vocabulary

tens   a place value
of a number

ones   a place value
of a number

standard form   the
usual way of writing a
number that shows only
its digits, no words
537   89   1,642

23
This number has
2 tens and **3** ones.

The number 35 is
between 34 and 36.

Write the missing number.
57, 58, 59, _____, 61, 62

| 51 | 52 | 53 | 54 | 55 | 56 | 57 | 58 | 59 | 60 |
| 61 | 62 | 63 | 64 | 65 | 66 | 67 | 68 | 69 | 70 |

Step 1   Begin at the first number. 57
Step 2   The ones digit goes up by 1.
          57, 58, 59
Step 3   What number is after 59? 60

Answer   The missing number is 60.

Step-by-Step Practice

Write the missing number.
21, 22, 23, _____, 25, 26

| 21 | 22 | 23 | 24 | 25 | 26 | 27 | 28 | 29 | 30 |

Step 1   Begin at the first number. _____
Step 2   The ones digit goes up by 1.

          _____, _____, _____

Step 3   What number is after 23? _____

Answer   The missing number is _____.

Name _____

 Guided Practice

Write each missing number.

1

71	72	73	74	75	76	77	78	79	80

2 84, 85, 86, _87_, 88   **3** _____, 95, 96, 97, 98

---

4 Match each number with the correct place values.

75   43   21   15

7 tens 5 ones	2 tens 1 one	1 ten 5 ones	4 tens 3 ones

Problem-Solving Practice

5 The mystery number comes just after 65 and just before 67. What is the mystery number?

Understand Underline key words.

Plan Count by ones.

Solve Start at 60 and count on by ones.

_____, _____, _____, _____, _____, _____, _____, _____

The mystery number is _____.

Check Count back from 70. Is your answer the same?

GO on

▶ Practice on Your Own

Write each missing number.

6

51	52	53	54	55	*56*	57	58	59	60

7

31	32	33	34	35	36	37	38	39	*40*
41	42	43	*44*	45	46	47	48	49	50

8 _*18*_, 19, 20, 21, 22   **9** 77, 78, _*79*_, 80, 81

10 66, 67, 68, 69, _*70*_   **11** 91, 92, _*93*_, 94, 95

Complete each place value.

12 37 = _____ tens _____ ones   **13** 48 = _____ tens _____ ones

14 16 = _____ ten _____ ones   **15** 91 = _____ tens _____ one

16 ◀ **WRITING IN** ▶**MATH**   Emilio says his house number has 3 tens and 8 ones. Is Emilio correct? Explain.

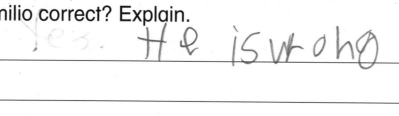 _Yes. He is wrong_

Vocabulary Check   Complete.

17 The number 70 has 7 _*67*_.

STOP

Name _____

Expanded Form

Key Concept

Numbers can be written in different forms.

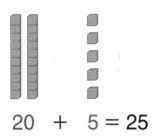

20 + 5 = 25

The picture shows 2 tens rods and 5 ones cubes. The picture shows 25 blocks in all.

standard form	25
expanded form	20 + 5
word form	twenty-five

Vocabulary

standard form   the usual way of writing a number that shows only its digits, no words

expanded form   the way of writing a number as a sum that shows the value of each digit

word form   the way of writing a number in words, no numbers

Use a hyphen to separate two or more tens from the ones. twenty-one
↑

Use an addition sign to show the tens plus the ones. 21 = 20 + 1
↑

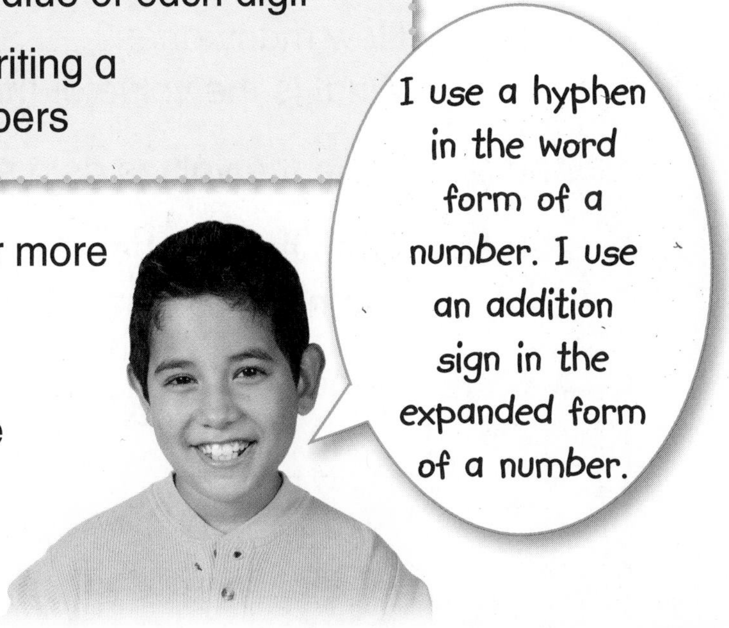

I use a hyphen in the word form of a number. I use an addition sign in the expanded form of a number.

Example

Complete the table.

Step 1   How many tens? 3
             What is the value of the tens? 30

Step 2   How many ones? 7
             What is the value of the ones? 7

Step 3   Write the values as a sum. 30 + 7

Step 4   Write the number
             and number name
             of the values. 37,
             thirty-seven

standard form	37
expanded form	30 + 7
word form	thirty-seven

Step-by-Step Practice

Complete the table.

Step 1   How many tens? _____
             What is the value of the tens? _____

Step 2   How many ones? _____
             What is the value of the ones? _____

Step 3   Write the values as a sum. _____

Step 4   Write the number
             and number name
             of the values. _____,

             _____

standard form	
expanded form	
word form	

Name _____

 Guided Practice

Complete each table.

1

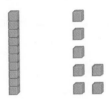

standard form	
expanded form	
word form	seventeen

2

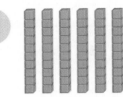

standard form	62
expanded form	
word form	

Problem-Solving Practice

3 I am greater than 50, but less than 60.
I have 5 tens and 4 ones. What number am I?

Understand Underline key words.

Plan Work backward.

Solve 5 tens = _____

4 ones = _____

_____ + _____ = _____

The number is _____.

Check Use place-value blocks to check your answer.

GO on

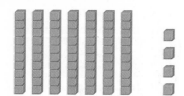 **Practice on Your Own**

Write each number in expanded form.

4 74 = _____ + _____

5 53 = _____ + _____

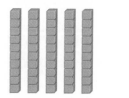

Complete each table.

6

standard form	92
expanded form	
word form	

7

standard form	
expanded form	
word form	forty-three

8

standard form	
expanded form	20 + 1
word form	

9

standard form	69
expanded form	
word form	

10 **WRITING IN ►MATH** Marcos wants to write 99 in expanded form. He writes 9 + 9. Is Marcos correct? Explain.

_____

_____

Vocabulary Check Complete.

11 The _____ form of 38 is 30 + 8.

STOP

128 one hundred twenty-eight

Name _____

Progress Check 1 (Lessons 4-1 and 4-2)

Write each missing number.

1

31	32	33	34		36	37	38	39	40

2

41	42		44	45	46	47	48	49	50

Complete each table.

3

standard form	
expanded form	
word form	sixty-seven

4

standard form	
expanded form	90 + 1
word form	

Write each number in standard form.

5 twenty-one _____

6 sixty-eight _____

Write each number in word form.

7 75 _____

8 86 _____

9 I am a number greater than 30 but less than 40.
I have 3 tens and 8 ones. What number am I?
How do you know?

_____

_____

Name _____

Connect the dots from 50 to 100 to find the answer.

Name _____

Round Two-Digit Numbers

Key Concept

You can **round** numbers to the nearest ten.
Look at 28 on the **number line**.

28 is between 20 and 30.
Is 28 closer to 20 or 30?

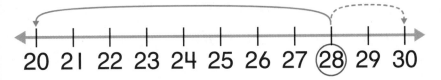

20 21 22 23 24 25 26 27 (28) 29 30

28 is 8 away from 20.
28 is 2 away from 30.
28 is closer to 30, so it rounds up to 30.

Vocabulary

round   to find the nearest value of a number based on a given place value
24 rounded to the nearest ten is 20.

number line   a line with number labels

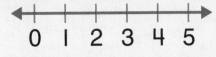

0 1 2 3 4 5

When a number has a 5 in the ones place, it rounds up. 25 rounds up to 30.

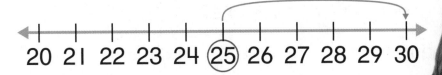

20 21 22 23 24 (25) 26 27 28 29 30

Example

There are 34 soup cans on a shelf.
About how many soup cans are on the shelf?
Round to the nearest ten.

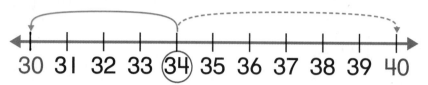

30 31 32 33 (34) 35 36 37 38 39 40

Step 1   34 is between 30 and 40. Circle 34.
Step 2   The nearest ten to the left of 34 is 30.
Step 3   The nearest ten to the right of 34 is 40.
Step 4   34 is closer to 30.

Answer   There are about 30 soup cans on the shelf.

Step-by-Step Practice

Mary bakes 67 cookies for a bake sale.
About how many cookies does Mary bake?
Round to the nearest ten.

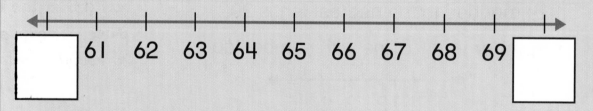

61 62 63 64 65 66 67 68 69

Step 1   67 is between _____ and _____. Circle 67.

Step 2   The nearest ten to the left of 67 is _____.

Step 3   The nearest ten to the right of 67 is _____.

Step 4   67 is closer to _____.

Answer   Mary bakes about _____ cookies.

Name _____

 Guided Practice

Use the number line to round each number.

1 73 rounds to _____

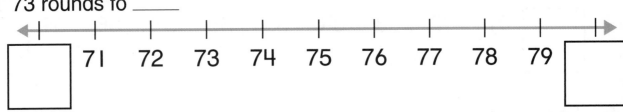

□ 71 72 73 74 75 76 77 78 79 □

2 45 rounds to _____

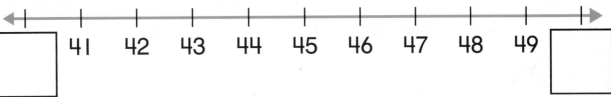

□ 41 42 43 44 45 46 47 48 49 □

Problem-Solving Practice

3 Simon has about 70 baseball cards when he rounds to the nearest ten. Could Simon have 62, 64, or 66 baseball cards?

Understand   Underline key words.

Plan   Count backward.

Solve   70, 69, 68, _____, _____, _____, _____, _____,

_____, _____, _____

_____ is closest to 70. It rounds to 70.

Simon could have _____ baseball cards.

Check   Use a number line.

GO on

▶ Practice on Your Own

Use the number line to round each number.

4 32 rounds to _____

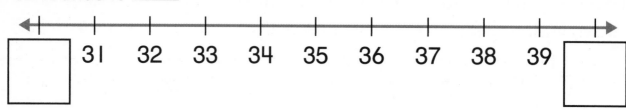

5 56 rounds to _____

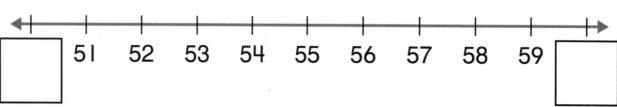

Fill in the blanks.

6 23 is between _____ and _____. 23 rounds to _____.

7 62 is between _____ and _____. 62 rounds to _____.

8 **WRITING IN ▶MATH** Halima has about 90 photos when she rounds to the nearest ten. Could Halima have 81, 84, or 88 photos? Explain.

_____

_____

Vocabulary Check Complete.

9 When you _____ a number, you find the nearest value of the number based on a given place value.

Copyright © Macmillan/McGraw-Hill • Glencoe, a division of The McGraw-Hill Companies, Inc.

STOP

Name _____

Whole Numbers Less Than 10,000

Key Concept

You can use base-ten blocks and digits to show numbers less than 10,000.

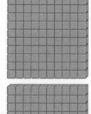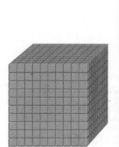

| 1 thousand | 3 hundreds | 4 tens | 6 ones |

You can write the number in a **place-value chart**. The number is 1,346.

thousands	hundreds	tens	ones
1	3	4	6

Vocabulary

thousands   a place value of a number

1,253

The 1 is in the thousands place.

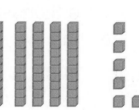
You separate the thousands digit and hundreds digit with a comma.

place-value chart   a chart that represents the place value of digits in a number

thousands	hundreds	tens	ones
1	2	5	3

Example

Write the number.

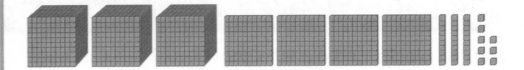

Step 1   Count. There are 3 thousands blocks.
Step 2   Count. There are 4 hundreds flats.
Step 3   Count. There are 3 tens rods.
Step 4   Count. There are 8 ones cubes.
Step 5   Complete the chart.

Answer   The number is 3,438.

thousands	hundreds	tens	ones
3	4	3	8

Step-by-Step Practice

Write the number.

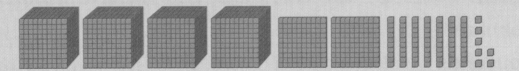

Step 1   Count. There are _____ thousands blocks.

Step 2   Count. There are _____ hundreds flats.

Step 3   Count. There are _____ tens rods.

Step 4   Count. There are _____ ones cubes.

Step 5   Complete the chart.

Answer   The number is _____.

thousands	hundreds	tens	ones

Name _____

 Guided Practice

Write each number.

1 _____

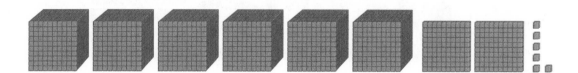

2 A number has 5 thousands, 2 tens, and 8 ones.

What is the number? _____

Problem-Solving Practice

3 Theo rides the bus to school.
The number of the bus he rides has
8 thousands, 6 hundreds, 5 tens, and 1 one.
What is the number of Theo's bus?

Understand   Underline key words.

Plan   Use a place-value chart.

Solve   Write each clue in the place-value chart.

thousands	hundreds	tens	ones

The number of Theo's bus is _____.

Check   Use base-ten blocks to model the number.

Practice on Your Own

4 Write the number.

_____

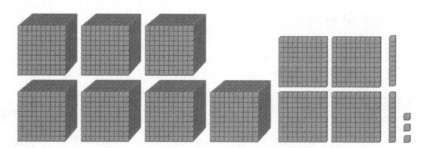

---

5 A number has 9 thousands,
4 hundreds, 5 tens, and 1 one.
What is the number?

thousands	hundreds	tens	ones

6

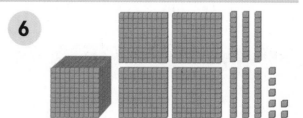

thousands	hundreds	tens	ones

---

7 **WRITING IN ►MATH** How many tens are in 4,307? Explain.

_____

_____

Vocabulary Check Complete.

8 In 3,867, the digit 3 is in the _____ place.

Name _____

Progress Check 2 (Lessons 4-3 and 4-4)

1 Use the number line to round the number.

77 rounds to _____.

```
[ ]   71   72   73   74   75   76   77   78   79   [ ]
```

Fill in the blanks.

2 81 is between _____ and _____. 81 rounds to _____.

3 95 is between _____ and _____. 95 rounds to _____.

---

4 Write the number.

_____

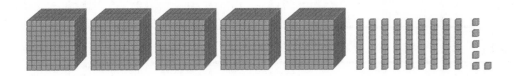

---

5 Complete the place-value chart.
A number has 8 thousands,
7 hundreds, and 3 ones.
What is the number?

thousands	hundreds	tens	ones

---

6 Alexis has about 50 stamps when she rounds to the nearest ten. Could Alexis have 43, 48, or 57 stamps?

Alexis could have _____ stamps.

Name _____

How to Play

1 Players 1 and 2 take turns rolling the number cubes to make a two-digit number.

2 Round each number to the nearest ten. Record the answers on the score card.

3 Circle the greater rounded number. The player with the greater rounded number earns one point. The first player to earn three points wins!

Materials
score card
number cube 1– 6
blank number cube
(label sides 4 – 9)

Ready to Round Game	Player 1		Player 2	
	number	Round to the Nearest 10	number	Round to the Nearest 10
Roll 1				
Roll 2				
Roll 3				
Roll 4				
Roll 5				
Score				

I rolled the numbers 4 and 8. I can make the number 84, which rounds to 80.

140 one hundred forty

Copyright © Macmillan/McGraw-Hill • Glencoe, a division of The McGraw-Hill Companies, Inc.

Name _____

Review

Copyright © Macmillan/McGraw-Hill, • Glencoe, a division of The McGraw-Hill Companies, Inc.

Vocabulary

Word Bank	Use the Word Bank to complete.
expanded form **round** **standard form** **word form**	**1** eighty-two _____ **2** 67 _____ **3** 26 → 30 _____ **4** 40 + 5 _____

 **Concepts**

Write each missing number.

5

71	72	73	74	75	76		78	79	80

6

81	82	83	84	85	86	87	88		90

Complete each table.

7

standard form	68
expanded form	
word form	

8

standard form	
expanded form	30 + 5
word form	

GO on

Write each number in standard form.

9 seventy-three   **10** twenty-nine   **11** forty-eighty

_____   _____   _____

Write each number in expanded form.

12 75 _____   **13** 86 _____

Fill in the blanks.

14 67 is between _____ and _____. 67 rounds to _____.

15 75 is between _____ and _____. 75 rounds to _____.

16 I am a number greater than 70 but less than 80.

I have 7 tens and 3 ones. What number am I? _____

Write each number in a place-value chart.

17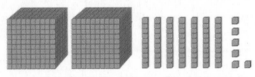

thousands	hundreds	tens	ones

18 A number has 7 thousands, 8 hundreds, 1 ten, and 4 ones. What is the number?

thousands	hundreds	tens	ones

19 Thomas has about 80 buttons when he rounds to the nearest ten. Could Thomas have 72, 75, or 86 buttons? Explain.

_____

_____

STOP

Review

Name _____

Chapter Test

Write each missing number.

1 46, 47, 48, _____, 50   **2** _____, 32, 33, 34, 35

3 73, _____, 75, 76, _____   **4** _____, 97, 98, 99, _____

Complete each table.

5

standard form	
expanded form	60 + 3
word form	

6

standard form	85
expanded form	
word form	

Fill in the blanks.

7 43 is between _____ and _____. 43 rounds to _____.

8 25 is between _____ and _____. 25 rounds to _____.

Write each number. Fill in the place-value chart.

9

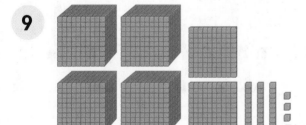

_____

thousands	hundreds	tens	ones

10 A number has 2 thousands, 3 tens, and 4 ones. What is the number?

_____

thousands	hundreds	tens	ones

GO on

11 Who is Correct?

Juanita and Matt write the number shown.

Juanita   Matt

Circle the correct answer. Explain.

_____

_____

Fill in the blanks.

12 The players line up by number. Which player is missing?

Player number _____ is missing.

13 I am a number greater than 80 but less than 90. I have 5 ones and I round to 90. What number am I? _____

14 Vickie has about 20 crayons when she rounds to the nearest ten. Could Vickie have 14, 22, or 28 crayons? Explain.

_____

_____

STOP

Chapter 4 Test

Name _____

Test Practice

Choose the correct answer.

1 What is the missing number?

40, 41, 42, _____, 44, 45

34   41   43   46
○   ○   ○   ○

2 What is the missing number?

93, 94, 95, _____, 97, 98

69   91   95   96
○   ○   ○   ○

3 Bernard uses base-ten blocks to model his favorite number. What is Bernard's favorite number?

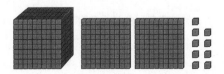

129   1,029   1,209   1,290
○   ○   ○   ○

4 Polly sold 48 boxes of cookies for a fundraiser. About how many boxes did Polly sell? Round to the nearest ten.

40   45   50   60
○   ○   ○   ○

5 What is 21 rounded to the nearest ten?

10   20   30   40
○   ○   ○   ○

6 Ryan wants to write the expanded form of the number 86 for his little brother. What should Ryan write?

○ 8 + 6   ○ 8 + 60

○ 80 + 6   ○ 80 + 60

GO ON

7 What is the standard form of 40 + 7?

47 74 407 704
○   ○   ○   ○

8 What number is shown?

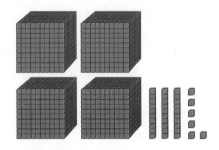

- ○ 463
- ○ 4,036
- ○ 4,306
- ○ 4,360

9 Arnaldo is counting his money. He has 50 + 9 pennies. What is this number in word form?

- ○ forty-nine   ○ fifty
- ○ fifty-nine   ○ ninety-five

10 What is the missing number?

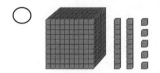

23, 24, 25, _____, 27, 28

20 22 26 29
○   ○   ○   ○

11 Which model shows 1,025?

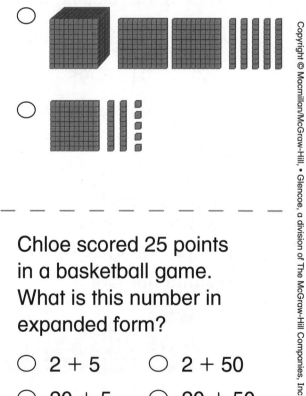

12 Chloe scored 25 points in a basketball game. What is this number in expanded form?

- ○ 2 + 5   ○ 2 + 50
- ○ 20 + 5   ○ 20 + 50

Home Connection

English

Dear Family,
Today our class started **Chapter 5, Fractions.** In this chapter, I will learn about fractions as parts of a whole and as parts of a set. I will also find fractions on a number line.

Love, _____

Spanish

Estimada familia:
Hoy en clase comenzamos el **Capítulo 5, Fracciones.** En este capítulo aprenderé sobre las fracciones como partes de un entero y como partes de un set. También aprenderé a encontrar fracciones en una recta numérica.

Cariños, _____

Help at Home

You can help your child learn about fractions. Have your child fold or cut a paper plate or another household item into equal parts. Have him or her identify different fractions of the whole.

Ayude en casa

Usted puede ayudar a su hijo(a) a aprender sobre las fracciones. Pida a su niño que doble y corte en partes iguales un plato de cartón u otra cosa que tenga en casa. Pídale que identifique las diferentes fracciones de un entero.

Math Online Take the chapter Get Ready quiz at macmillanmh.com.

Name _____

Get Ready

Count. Write how many.

 1

_____

2

_____

Write equal or unequal.

3

_____

4

_____

5

_____

6

_____

Draw a model for each number.

7 5

8 7

Complete each number line.

9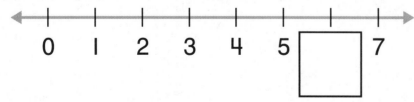

0   1   2   3   4   5   [   ]   7

10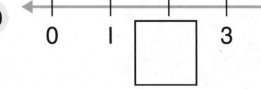

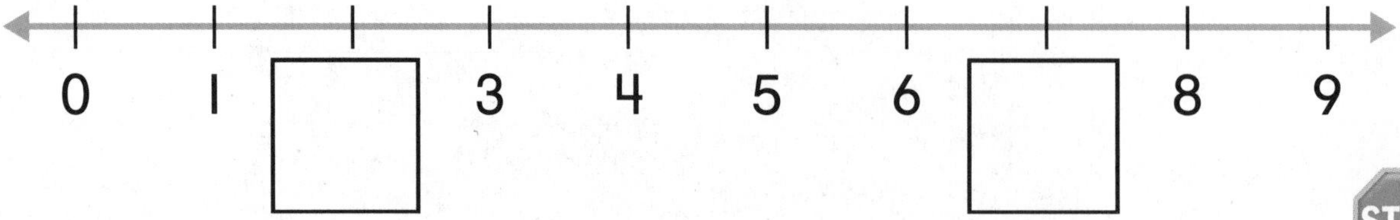

0   1   [   ]   3   4   5   6   [   ]   8   9

STOP

148   one hundred forty-eight

Name _____

Equal Parts

Key Concept

You can divide an object into parts.

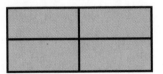   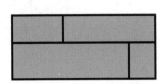

The orange rectangle is divided into 4 **equal parts**.

The blue rectangle is divided into 4 unequal parts.

Vocabulary

whole   the entire amount or object

equal parts   each part is the same size

This sandwich is cut into 2 equal parts.

You can fold a piece of paper to show equal parts.

Place one part on top of another to show the parts are equal.

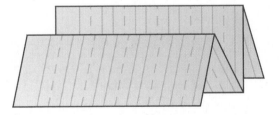

Example

Circle the figure that shows equal parts.
Write how many equal parts.

 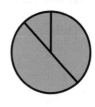

Step 1   Look at the parts of each figure.

Step 2   Circle the figure with equal parts.

Step 3   Count the number of equal parts. 1, 2, 3

Answer   There are 3 equal parts.

Step-by-Step Practice

Circle the figure that shows equal parts.
Write how many equal parts.

Step 1   Look at the parts of each figure.

Step 2   Circle the figure with equal parts.

Step 3   Count the number of equal parts.

____, ____, ____, ____

Answer   There are _____ equal parts.

Name _____

Circle each figure that shows equal parts.

How many equal parts are there?

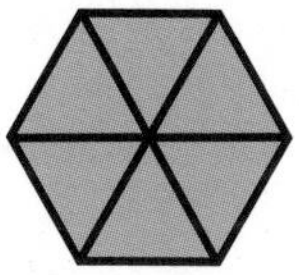

_____ equal parts

_____ equal parts

Problem-Solving Practice

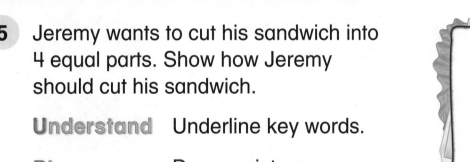

5 Jeremy wants to cut his sandwich into 4 equal parts. Show how Jeremy should cut his sandwich.

Understand	Underline key words.
Plan	Draw a picture.
Solve	Draw lines to show _____ equal parts.
Check	Look at the parts of the sandwich. Are they equal?

GO on

▶ Practice on Your Own

Circle each object that shows equal parts.

6 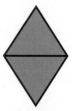    7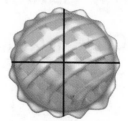

How many equal parts are there?

8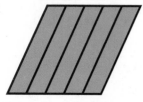

_____ equal parts

9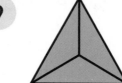

_____ equal parts

---

10   **WRITING IN ▶ MATH**   Mrs. Thomas wants
to cut a pie into 8 equal slices. How can
you show this?

_____

_____

_____

Vocabulary Check   Complete.

11   An object with same size parts is divided

into _____.

152   one hundred fifty-two

STOP

Name _____

One-Half, One-Third, and One-Fourth

Key Concept

You can use a **fraction** to name part of a whole.

one-half or $\frac{1}{2}$

1 out of 2 parts is blue.

one-third or $\frac{1}{3}$

1 out of 3 parts is red.

one-fourth or $\frac{1}{4}$

1 out of 4 parts is green.

Vocabulary

fraction   a number that represents part of a whole or part of a set

numerator   part of the fraction that tells how many equal parts are being used

$$\frac{5}{6} \longleftarrow \text{numerator}$$

denominator   part of the fraction that tells the total number of equal parts

$$\frac{5}{6} \longleftarrow \text{denominator}$$

The **numerator** tells the number of shaded parts.

$\frac{1}{2}$ ← The **denominator** tells the number of equal parts.

Example

Write the fraction that names the shaded part.

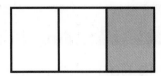

Step 1   Are the parts equal? yes

Step 2   The numerator tells the number of shaded parts.
             The numerator is 1.

Step 3   The denominator tells the number of equal parts.
             The denominator is 3.

Answer   The fraction is $\dfrac{1}{3}$.

Step-by-Step Practice

Write the fraction that names the shaded part.

Step 1   Are the parts equal? _____

Step 2   The numerator tells the number of shaded parts.

             The numerator is _____.

Step 3   The denominator tells the number of equal parts.

             The denominator is _____.

Answer   The fraction is $\dfrac{\square}{\square}$.

Name _____

 Guided Practice

Write the fraction that names each shaded part.

 1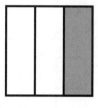

$$\frac{\square}{\square}$$

2

$$\frac{\square}{\square}$$

3

$$\frac{\square}{\square}$$

Problem-Solving Practice

4   Devon wants to color $\frac{1}{2}$ of a banner red.
    Show what will the banner look like.

Understand   Underline key words.

Plan   Use a model.

Solve   Divide the banner into

   _____ equal parts.

   Color _____ part red.

Check   Is the shaded part of the banner
   equal to the unshaded part?

GO on

Copyright © Macmillan/McGraw-Hill • Glencoe, a division of The McGraw-Hill Companies, Inc.

▶ Practice on Your Own

Write the fraction that names each shaded part.

5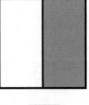

☐
—
☐

6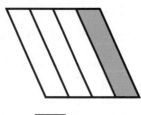

☐
—
☐

7

☐
—
☐

8

☐
—
☐

9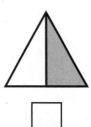

☐
—
☐

10

☐
—
☐

---

11 **WRITING IN ►MATH** Caitlyn is making a sign.

She wants to shade $\frac{1}{3}$ of the sign blue.

Show what this will look like. Explain.

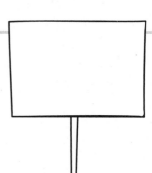

_____

_____

Vocabulary Check Complete.

12 A _____ is a number that represents part
of a whole.

STOP

156 one hundred fifty-six

Name _____

Progress Check 1 (Lessons 5-1 and 5-2)

1 Circle the figure that shows equal parts.

  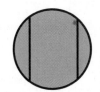

How many equal parts are there?

2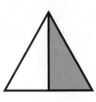

_____ equal parts

3

_____ equal parts

Write the fraction that names each shaded part.

4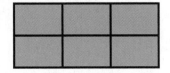

$\dfrac{\Box}{\Box}$

5

$\dfrac{\Box}{\Box}$

6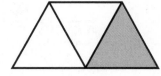

$\dfrac{\Box}{\Box}$

7 **WRITING IN ▶MATH** Kofi cut a large round cookie into 3 equal parts and ate 1 part. How much of the cookie did Kofi eat?

Name _____

The hot air balloon is missing its ropes!
Draw a line from each shape to the fraction that
matches the shaded parts.

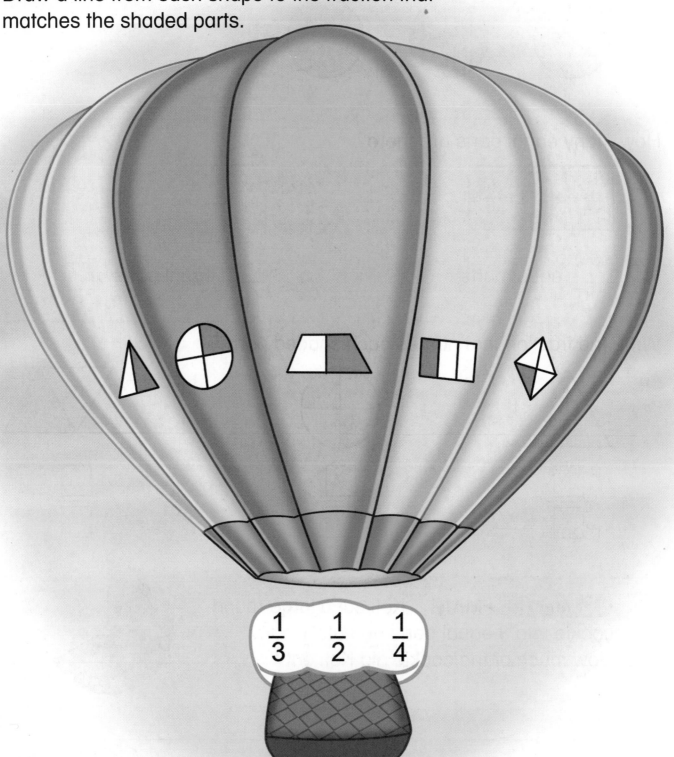

Name _____

Parts of a Whole

Key Concept

Fractions can name more than one equal part of a whole.

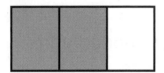

two-thirds or $\dfrac{2}{3}$ ← **numerator**

← **denominator**

The denominator shows that there are 3 equal parts.

The numerator shows that 2 of the 3 parts are shaded.

two-sixths or $\dfrac{2}{6}$ ← numerator

← denominator

There are 6 equal parts.
2 of the 6 parts are shaded.
4 of the 6 parts are unshaded.

Vocabulary

fraction   a number that represents part of a whole or part of a set

numerator   part of the fraction that tells how many equal parts are being used

$$\dfrac{5}{6} \longleftarrow \text{numerator}$$

denominator   part of the fraction that tells the total number of equal parts

$$\dfrac{5}{6} \longleftarrow \text{denominator}$$

Write the fraction that names the shaded part.

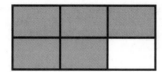

Step 1   Are the parts equal? yes

Step 2   The **numerator** tells the number of shaded parts.
          The numerator is 5.

Step 3   The **denominator** tells the number of equal parts.
          The denominator is 6.

Answer   The fraction is $\dfrac{5}{6}$.

Step-by-Step Practice

Write the fraction that names the shaded part.

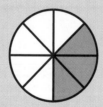

Step 1   Are the parts equal? _____

Step 2   The numerator tells the number of shaded parts.

          The numerator is _____.

Step 3   The denominator tells the number of equal parts.

          The denominator is _____.

Answer   The fraction is $\dfrac{\Box}{\Box}$.

Name _____

Guided Practice

Write the fraction that names each shaded part.

1

$$\frac{\square}{4}$$

2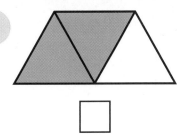

$$\frac{\square}{3}$$

3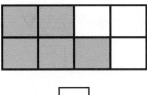

$$\frac{\square}{8}$$

Problem-Solving Practice

4 Sonia and three friends want to share a pizza.
Which pizza can the girls share so that each girl gets 1 slice?

Understand   Underline key words.

Plan   Guess and check.

Solve   There are _____ girls in all.

Which pizza has _____ slices?
Circle the pizza the girls can share equally.

Check   Shade 1 part of the pizza you circled
for each girl. Is there 1 part for each girl?

GO on

▶ Practice on Your Own

Write the fraction that names each shaded part.

5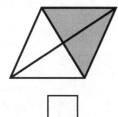

□/4

6

□/3

7

□/5

8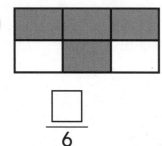

□/6

9

□/9

10

□/8

---

11 ✎ WRITING IN ▶MATH   Nashoba and two friends ate a small apple pie. Each boy ate an equal number of slices. Which pie could the boys have eaten? Explain.

_____

_____

Vocabulary Check   Complete.

12 The _____ is the part of the fraction that tells the total number of equal parts.

STOP

162   one hundred sixty-two

Name _____

Parts of a Set

Key Concept

Fractions can name part of a group or set.

What fraction of the buttons are red?

   I red button
4 buttons in the set

The fraction $\frac{1}{4}$ tells how many buttons are red.

What fraction of the counters are yellow?

3 yellow counters
5 counters in the set

The fraction $\frac{3}{5}$ tells how many counters are yellow.

Vocabulary

numerator   part of the fraction that tells how many equal parts are being used

$$\frac{5}{6} \longleftarrow \text{numerator}$$

denominator   part of the fraction that tells the total number of equal parts

$$\frac{5}{6} \longleftarrow \text{denominator}$$

The fraction $\frac{2}{7}$ tells how many leaves are green.

Name the fraction of orange leaves.

Step 1	There are 3 orange leaves.
Step 2	There are 4 leaves in the set.
Step 3	In the set, 3 out of 4 leaves are orange.

Answer The fraction is $\frac{3}{4}$.

Step-by-Step Practice

Name the fraction of red flowers.

Step 1	There are _____ red flowers.
Step 2	There are _____ flowers in the set.
Step 3	In the set, _____ out of _____ flowers are red.

Answer The fraction is $\frac{\square}{\square}$.

Name _____

Guided Practice

Name each fraction.

1

$\dfrac{\Box}{7}$ of the fruit are pears.

2

$\dfrac{\Box}{3}$ of the fruit are oranges.

Problem-Solving Practice

3 Molly's mom buys 8 apples at the store. She buys 4 red apples. Color the part of the apples that are red. Name the fraction.

Understand Underline key words.

Plan Color a picture.

Solve Color _____ out of _____ apples red.

$\dfrac{\Box}{\Box}$ of the apples are red.

Check Count the number of red apples. This is the numerator. Count the number of apples in the set. This is the denominator.

GO on

▶ Practice on Your Own

Name each fraction.

4

$\dfrac{\square}{4}$ of the animals are giraffes.

5

$\dfrac{\square}{3}$ of the birds are blue.

6

$\dfrac{\square}{6}$ of the dogs have red collars.

7

$\dfrac{\square}{7}$ of the farm animals are cows.

8 **WRITING IN ▶MATH**   Rafi has 4 cats. Of the cats, 1 is black. What fraction of the cats are black? How can you show this? Explain.

_____

_____

Vocabulary Check   Complete.

9 The _____ is the part of the fraction that tells how many equal parts are being used.

STOP

Name _____

Progress Check 2 (Lessons 5-3 and 5-4)

Write the fraction that names each shaded part.

1   $\dfrac{\Box}{\Box}$

2   $\dfrac{\Box}{\Box}$

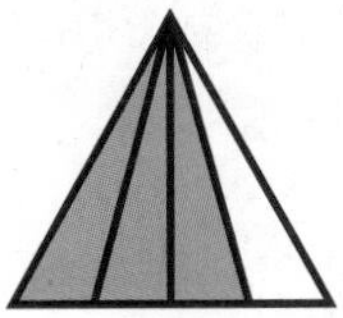

3   $\dfrac{\Box}{\Box}$

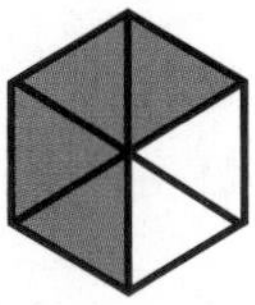

4   $\dfrac{\Box}{\Box}$

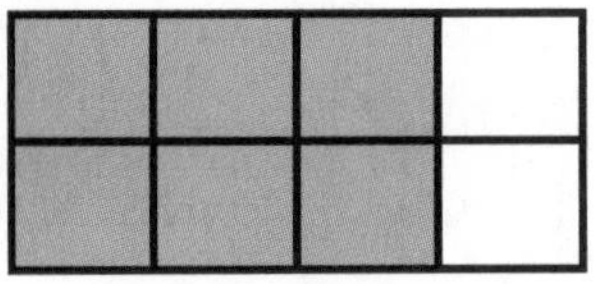

Name each fraction.

5 Jennifer bought 2 balls and 3 bones for her dog.
What fraction of the items are balls?

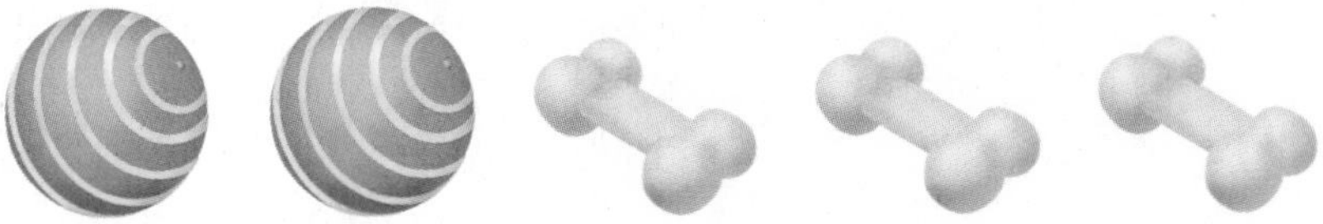

In the set, $\dfrac{\Box}{\Box}$ of the items are balls.

6 Tony bought 1 bed and 2 bowls for his dog.
What fraction of the items are bowls?

In the set, $\dfrac{\Box}{\Box}$ of the items are bowls.

Name _____

Complete the fractions to name each shaded part. Place your answers in order in the boxes below. Use the key to solve the riddle.

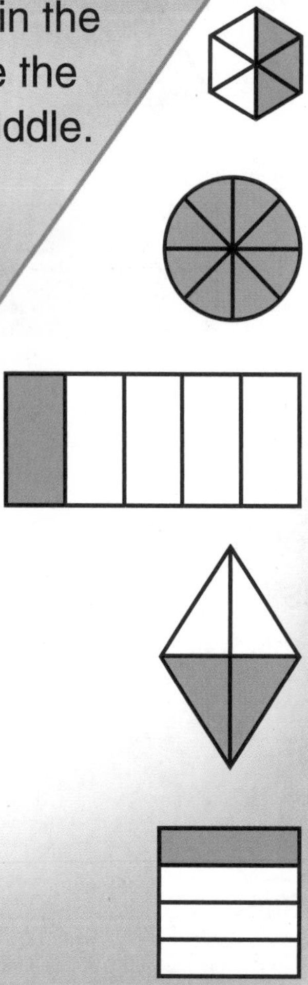

 $\dfrac{}{6}$

 $\dfrac{8}{}$

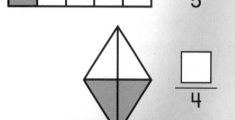 $\dfrac{}{5}$

$\dfrac{}{4}$

 $\dfrac{1}{}$

Key

1 = I	2 = L
3 = S	4 = N
8 = A	

I am ☐ ☐ ☐ ☐ I ☐ G through fractions!

168 one hundred sixty-eight

Model Fractions

Key Concept

You can model parts of a whole by drawing a picture.

$\dfrac{3}{4}$ ⟶ number of parts shaded
⟶ total number of parts in the whole

The **fraction** $\dfrac{3}{4}$ is modeled by the picture.

You can model parts of a set using counters.

$\dfrac{3}{4}$ ⟶ number of yellow counters
⟶ total number of counters

The fraction $\dfrac{3}{4}$ is modeled by the counters.

Vocabulary

fraction   a number that represents part of a whole or part of a set

numerator   part of the fraction that tells how many equal parts are being used

$$\dfrac{5}{6} \longleftarrow \text{numerator}$$

denominator   part of the fraction that tells the total number of equal parts

$$\dfrac{5}{6} \longleftarrow \text{denominator}$$

Example

Mercedes ate $\frac{2}{6}$ of a chicken pot pie.

Draw a picture to model the fraction.

Step 1   Draw 1 circle to represent the whole pie.

Step 2   Explain $\frac{2}{6}$.

$\frac{2}{6}$ $\longrightarrow$ number of parts eaten
$\phantom{\frac{2}{6}}$ $\longrightarrow$ total number of parts in the whole

Step 3   Divide the pie into 6 equal parts.

Step 4   Shade 2 parts.

Answer

Step-by-Step Practice

Alan ate $\frac{4}{9}$ of the berries.

Draw a picture to model the fraction.

Step 1   Draw _____ circles to represent the berries.

Step 2   Explain $\frac{4}{9}$.

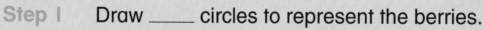

$\longrightarrow$ number of berries eaten

$\longrightarrow$ total number of berries

Step 3   Shade _____ circles.

Answer

Name ______________________________

Guided Practice

Draw a picture to model each fraction.

1 $\frac{2}{3}$ of a circle is shaded.

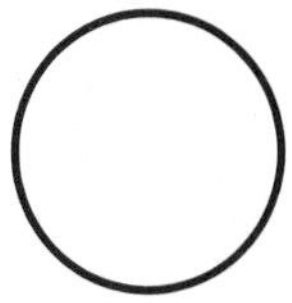

2 $\frac{4}{5}$ of a square is shaded.

3 $\frac{3}{8}$ of the stamps are animals.

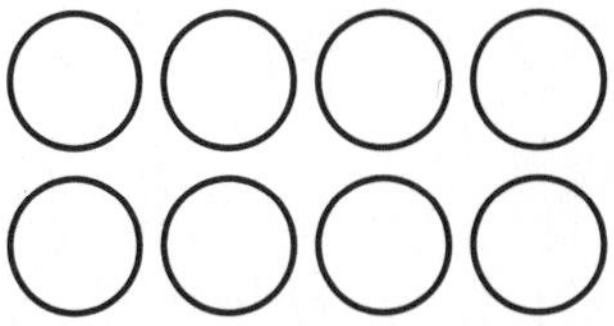

Problem-Solving Practice

4 Ama has 7 stickers in her sticker book. $\frac{3}{7}$ of the stickers are yellow. The other stickers are red. Show Ama's stickers.

Understand Underline key words.

Plan Use a model.

Solve Model the fraction using counters.

There are ____ stickers in all.

There are ____ yellow stickers.

There are ____ red stickers.

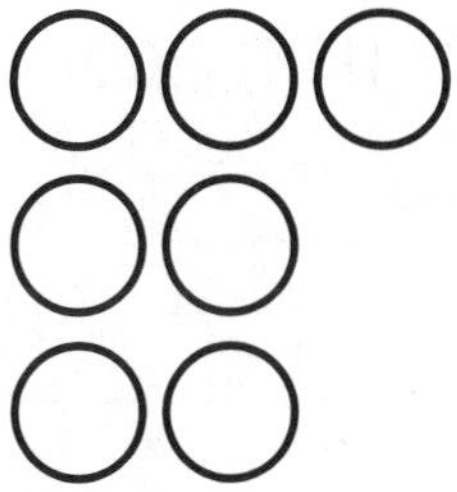

Check How many yellow counters did you use?
How many counters did you use in all?

GO on

▶ Practice on Your Own

Draw a picture to model each fraction.

5 $\dfrac{1}{3}$

6 $\dfrac{2}{4}$

Draw circles to model each fraction.

7 $\dfrac{3}{6}$

8 $\dfrac{1}{4}$

9 $\dfrac{2}{5}$

10 $\dfrac{6}{6}$

11 **WRITING IN ▶MATH** George has 10 baseball cards. His favorite player is shown on $\dfrac{6}{10}$ of the cards. How can you model this fraction using counters?

_____

_____

_____

Vocabulary Check Complete.

12 The _____ is the number above the bar in a fraction.

STOP

172 one hundred seventy-two

Name _____

Fractions on a Number Line

Key Concept

You can show **fractions** on a **number line**.

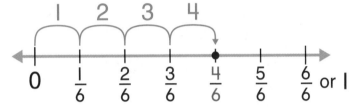

The space between the tick marks is $\frac{1}{6}$.

$\frac{4}{6}$ is 4 tick marks to the right of 0.

Copyright © Macmillan/McGraw-Hill, • Glencoe, a division of The McGraw-Hill Companies, Inc.

Vocabulary

fraction   a number that represents part of a whole or part of a set

number line   a line with number labels

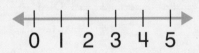

Count the tick marks between 0 and 1. Each tick mark shows the numerator. The denominator stays the same.

Bernice has a ribbon that is $\frac{3}{4}$-yard long.

Find the fraction on the number line.

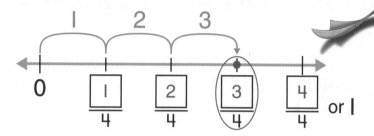

Step 1   Start at 0. Count each jump until you reach 3.

Step 2   Complete each fraction on the number line.

Step 3   Circle $\frac{3}{4}$ on the number line.

Step-by-Step Practice

Lee needs a $\frac{7}{8}$-inch toothpick for a craft project.

Find the fraction on the number line.

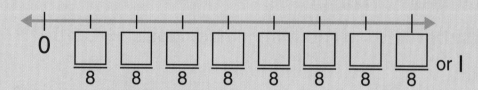

Step 1   Start at 0. Count each jump until you reach _____.

Step 2   Complete each fraction on the number line.

Step 3   Circle $\frac{7}{8}$ on the number line.

Name _____

Complete each fraction on the number line.
Circle the given fraction.

1 $\frac{3}{5}$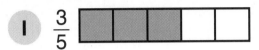

2 $\frac{2}{6}$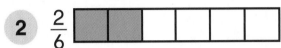

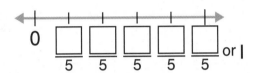

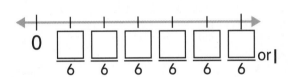

Problem-Solving Practice

3 Miguel has 4 hats. Of his hats, 2 are baseball caps. What fraction of Miguel's hats are baseball caps?

Understand Underline key words.

Plan Use a number line.

Solve Start at 0. Count until you reach _____.

☐
—— of Miguel's hats are baseball caps.
☐

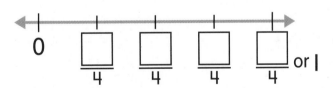

Check Draw a picture. Draw a circle with 4 equal parts. Shade 2 parts. What fraction is modeled?

GO on

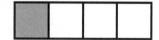

Practice on Your Own

Complete each fraction on the number line.
Circle the given fraction.

4

$$\frac{}{4} \quad \frac{}{4} \quad \frac{}{4} \quad \frac{}{4} \text{ or } 1$$

5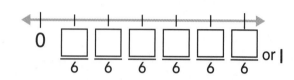

$$\frac{}{6} \quad \frac{}{6} \quad \frac{}{6} \quad \frac{}{6} \quad \frac{}{6} \quad \frac{}{6} \text{ or } 1$$

6 $\frac{2}{5}$

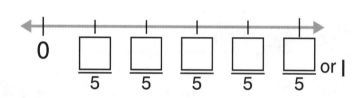

$$\frac{}{5} \quad \frac{}{5} \quad \frac{}{5} \quad \frac{}{5} \quad \frac{}{5} \text{ or } 1$$

7   **WRITING IN ►MATH**   Desta sees eight frogs.
Five are on lily pads. What fraction of the frogs
are on lily pads? How can you use a number line
to show your answer? Explain.

_____

_____

Vocabulary Check   Complete.

8   A _____ is a line with number labels.
It is divided by tick marks into equal parts.

STOP

Name _____

Progress Check 3 (Lessons 5-5 and 5-6)

Draw a picture to model each fraction.

1 $\frac{1}{3}$

2 $\frac{6}{8}$

Draw circles to model each fraction.

3 $\frac{3}{4}$

4 $\frac{2}{5}$

Complete each fraction on the number line.
Circle the given fraction.

5 $\frac{5}{6}$

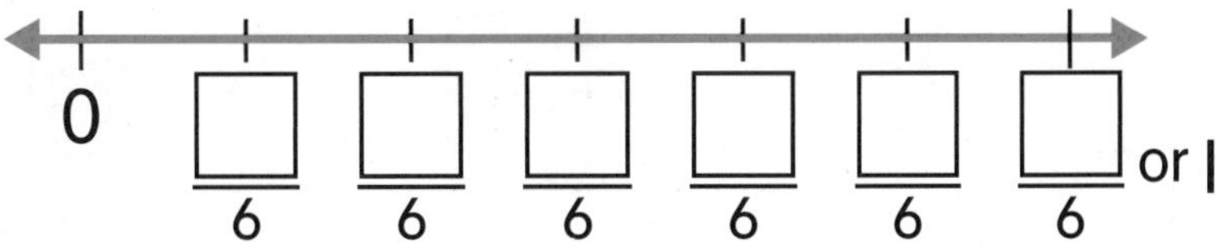

0 [ ]/6 [ ]/6 [ ]/6 [ ]/6 [ ]/6 [ ]/6 or 1

6 $\frac{4}{7}$

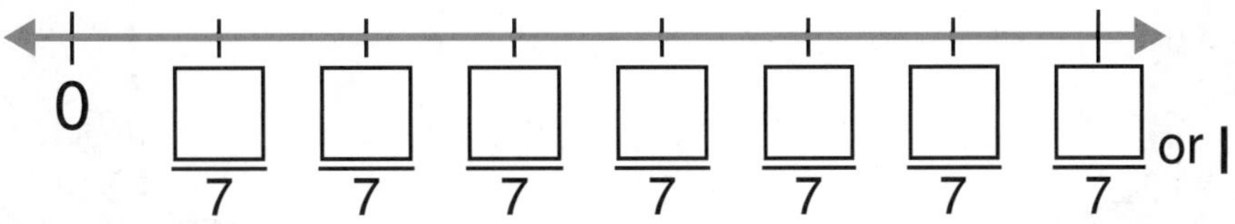

0 [ ]/7 [ ]/7 [ ]/7 [ ]/7 [ ]/7 [ ]/7 [ ]/7 or 1

7 Maggie ate $\frac{2}{8}$ of a pizza. Draw a picture
to model the amount Maggie ate.

Name ______________________________

«« Replay **Fraction Blast Off**

Draw a line to match each fraction with the correct model.

$\frac{3}{5}$

$\frac{1}{5}$

$\frac{1}{2}$

$\frac{4}{6}$

$\frac{2}{4}$

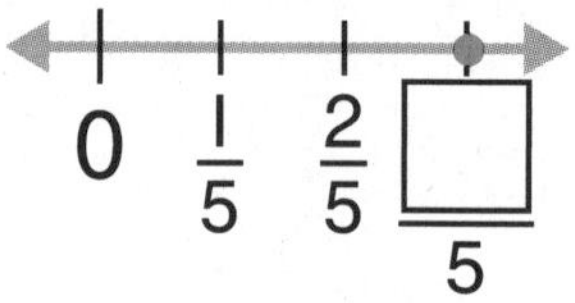

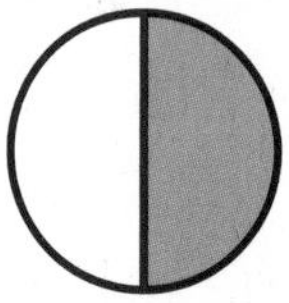

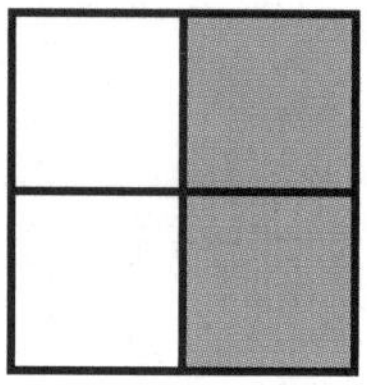

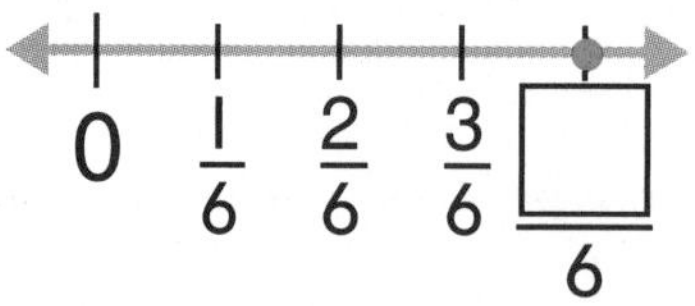

5 4 3 2 1 . . . **Blast off!**

Name _____

Review

Vocabulary

Word Bank

denominator

equal parts

number line

numerator

Use the Word Bank to complete.

1 $\dfrac{2}{5}$ ◄·· _____

2  _____

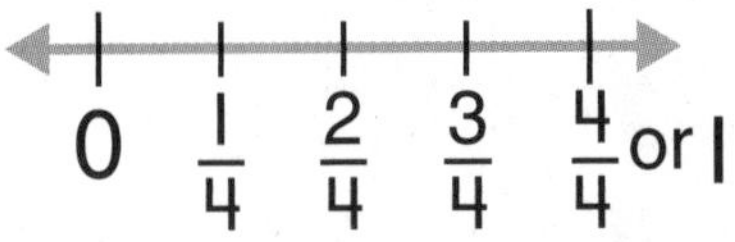

3 $\dfrac{1}{4}$ ◄····································· _____

4 _____

 ## Concepts

5 Circle the figure that shows equal parts.

      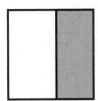

6 Draw circles to model the fraction.

$\dfrac{2}{4}$

GO on

Write the fraction that names each shaded part.

7 $\dfrac{\Box}{2}$

8  $\dfrac{\Box}{\Box}$

9 $\dfrac{\Box}{\Box}$

Name each fraction.

10  $\dfrac{2}{\Box}$ of the shirts are green.

11 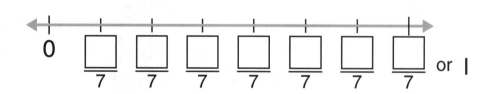 $\dfrac{\Box}{\Box}$ of the pants are blue.

Complete each fraction on the number line.
Circle the given fraction.

12 $\dfrac{4}{7}$

0   $\dfrac{\Box}{7}$ $\dfrac{\Box}{7}$ $\dfrac{\Box}{7}$ $\dfrac{\Box}{7}$ $\dfrac{\Box}{7}$ $\dfrac{\Box}{7}$ $\dfrac{\Box}{7}$ or 1

13 $\dfrac{3}{8}$

0   $\dfrac{\Box}{8}$ $\dfrac{\Box}{8}$ $\dfrac{\Box}{8}$ $\dfrac{\Box}{8}$ $\dfrac{\Box}{8}$ $\dfrac{\Box}{8}$ $\dfrac{\Box}{8}$ $\dfrac{\Box}{8}$ or 1

STOP

Review

Name _____

Chapter Test

How many equal parts are there?

1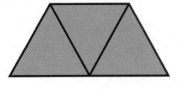

_____ equal parts

2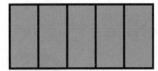

_____ equal parts

Write the fraction that names each shaded part.

3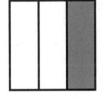

$\dfrac{\square}{\square}$

4

$\dfrac{\square}{\square}$

5

$\dfrac{\square}{\square}$

6 Name the fraction.

     $\dfrac{\square}{\square}$ of the counters are red.

Draw a picture to model each fraction.

7 $\dfrac{3}{7}$

8 $\dfrac{2}{6}$

GO on

9 Who is Correct?

Esi and Ethan write the fraction shown on the number line.

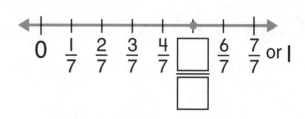

Esi

The number line shows $\frac{5}{7}$.

Ethan

The number line shows $\frac{5}{6}$.

Circle the correct answer. Explain.

_____

_____

---

10 Samantha ate $\frac{2}{3}$ of a sub sandwich. The sandwich was cut into 3 equal pieces. How many pieces did Samantha eat?

_____ pieces

11 In a bouquet of 8 flowers, $\frac{6}{8}$ are yellow.

Draw circles to model this fraction.

12 Ricardo has 4 red connecting cubes and 2 blue connecting cubes. What fraction of Ricardo's cubes are red? Show your answer on the number line.

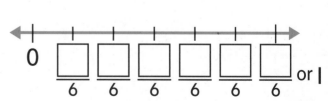

STOP

Name _____

Test Practice

Choose the correct answer.

1 Which figure shows 4 equal parts?

○   ○   ○   ○

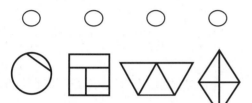

2 One-third of Jay's books are new. Which fraction shows one-third?

○   ○   ○   ○

$\dfrac{1}{3}$ $\dfrac{1}{2}$ $\dfrac{2}{3}$ $\dfrac{3}{1}$

3 What fraction names the shaded part?

○   ○   ○   ○

$\dfrac{1}{4}$ $\dfrac{1}{3}$ $\dfrac{1}{2}$ $\dfrac{2}{3}$

4 Which model shows $\dfrac{1}{4}$?

○   ○   ○   ○

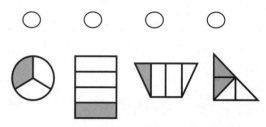

5 Three out of four students ride the bus. Which fraction shows 3 out of 4?

○   ○   ○   ○

$\dfrac{3}{4}$ $\dfrac{4}{4}$ $\dfrac{4}{3}$ $\dfrac{3}{1}$

6 Karl has 3 dimes and 2 nickels. What fraction of his coins are dimes?

○   ○   ○   ○

$\dfrac{1}{5}$ $\dfrac{2}{5}$ $\dfrac{3}{5}$ $\dfrac{2}{3}$

GO ON

7 What fraction names the shaded part?

○ $\dfrac{2}{8}$ ○ $\dfrac{3}{8}$ ○ $\dfrac{5}{8}$ ○ $\dfrac{6}{8}$

8 Which fraction is shown on the number line?

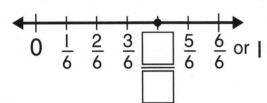

0 $\dfrac{1}{6}$ $\dfrac{2}{6}$ $\dfrac{3}{6}$ ☐ $\dfrac{5}{6}$ $\dfrac{6}{6}$ or 1

○ $\dfrac{2}{6}$ ○ $\dfrac{3}{6}$ ○ $\dfrac{4}{6}$ ○ $\dfrac{4}{5}$

9 Seven-eighths of Mrs. Riggle's plants are flowers. Which fraction shows seven-eighths?

○ $\dfrac{1}{8}$ ○ $\dfrac{1}{7}$ ○ $\dfrac{7}{8}$ ○ $\dfrac{8}{7}$

10 Which fraction shows the number of baseballs?

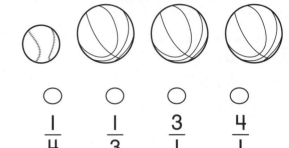

○ $\dfrac{1}{4}$ ○ $\dfrac{1}{3}$ ○ $\dfrac{3}{1}$ ○ $\dfrac{4}{1}$

11 Richard draws and shades a square to model a fraction. What fraction does he model?

○ $\dfrac{1}{5}$ ○ $\dfrac{2}{5}$ ○ $\dfrac{3}{5}$ ○ $\dfrac{5}{2}$

12 In the set, what fraction shows the bottles of glue?

○ $\dfrac{2}{5}$ ○ $\dfrac{1}{4}$ ○ $\dfrac{4}{5}$ ○ $\dfrac{4}{1}$

STOP

Test Practice

Chapter 6 Fraction Equivalence

Home Connection

English

Dear Family,
Today our class started **Chapter 6, Fraction Equivalence.** In this chapter, I will learn about equivalent fractions. I will also learn how to compare fractions and use fractions in measurement.

Love, _____

Spanish

Estimada familia:
Hoy en clase comenzamos el **Capítulo 6, Equivalencia de fracciones.** En este capítulo aprenderé sobre las fracciones equivalentes. También aprenderé cómo comparar fracciones y a usar fracciones en la medición.

Cariños, _____

Help at Home

You can compare fractions with your child at home. Have your child identify objects that show equal parts. Then, use fractions to name the objects. Ask your child to compare the fractions using the terms greater than and less than.

Ayude en casa

Usted puede practicar comparando fracciones con su hijo(a) en casa. Haga que identifique objetos que muestren partes iguales. Luego, use fracciones para nombrar los objetos. Pídale que compare las fracciones usando los términos mayor que y menor que.

Math Online Take the chapter Get Ready quiz at macmillanmh.com.

Name _____

Get Ready

Write the fraction for each shaded part.

1

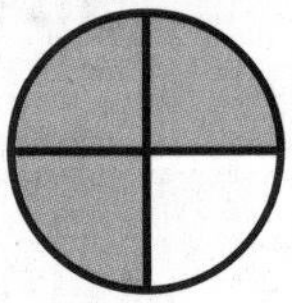

$$\frac{\Box}{\Box}$$

2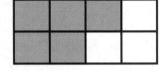

$$\frac{\Box}{\Box}$$

3

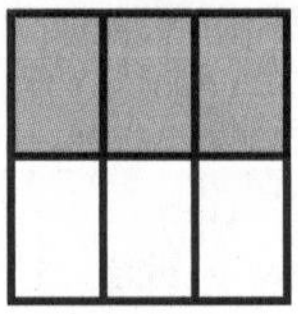

$$\frac{\Box}{\Box}$$

Compare. Write >, <, or =.

4 8 $\bigcirc$ 3   5 4 $\bigcirc$ 4   6 5 $\bigcirc$ 9   7 2 $\bigcirc$ 1

Measure to the nearest inch.

8 _____ inches

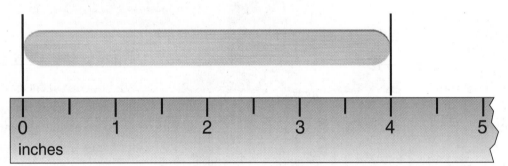

9 _____ inches

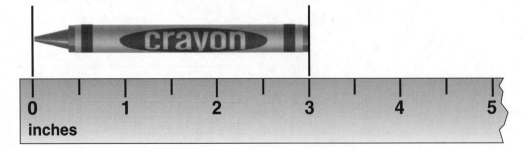

STOP

186 **one hundred eighty-six**

Fractions Equal to 1

Key Concept

Name a **fraction** for one whole.

Look at the wheel. There are 6 equal parts.

All 6 parts make up 1 wheel.

So, $\frac{6}{6}$ equals 1 whole wheel.

$\frac{6}{6} = 1$

Vocabulary

fraction   a number that represents part of a whole or part of a set

$$\left(\frac{1}{2}\right), \left(\frac{1}{3}\right), \left(\frac{1}{4}\right), \left(\frac{3}{4}\right)$$

| 1 | $\frac{1}{2}$ | $\frac{1}{2}$ |

| $\frac{1}{4}$ | $\frac{1}{4}$ |
| $\frac{1}{4}$ | $\frac{1}{4}$ |

A fraction with the same numerator and denominator is equal to 1, or one whole.

There are 9 equal window panes.
All 9 panes make up 1 whole window.
$\frac{9}{9} = 1$

Example

Circle the fraction that equals 1.

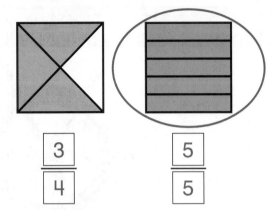

$$\dfrac{3}{4} \qquad \dfrac{5}{5}$$

Step 1   Write each fraction.

Step 2   Circle the fraction with the same numerator and denominator.

Answer $\dfrac{5}{5} = 1$

Step-by-Step Practice

Circle the fraction that equals 1.

Step 1   Write each fraction.

Step 2   Circle the fraction with the same numerator and denominator.

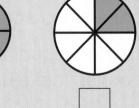

Answer $\dfrac{\boxed{}}{\boxed{}} = \underline{}$

Name _____

Name each fraction. Circle each fraction that equals 1.

1

$$\frac{\Box}{\Box}$$

2

$$\frac{\Box}{\Box}$$

3

$$\frac{\Box}{\Box}$$

Circle each fraction that equals 1.

4 $\dfrac{3}{3}$ $\dfrac{3}{5}$ $\dfrac{1}{3}$

5 $\dfrac{7}{8}$ $\dfrac{8}{10}$ $\dfrac{8}{8}$

Problem-Solving Practice

6 A birthday cake is cut into 7 pieces. What fraction shows the whole cake?

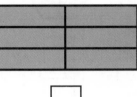

Understand Underline key words.

Plan Draw a picture.

Solve "Cut" the circle into 7 pieces. Shade the whole circle.

$$\frac{\Box}{\Box}$$ ← number of shaded pieces

← total number of pieces

Check Does the fraction have the same numerator and denominator?

GO on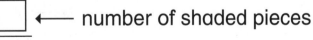

▶ Practice on Your Own

Name each fraction. Circle each fraction that equals I.

7

□/□

8

□/□

9

□/□

Circle each fraction that equals I.

10 $\frac{6}{7}$ $\frac{7}{7}$ $\frac{7}{8}$

11 $\frac{10}{14}$ $\frac{14}{14}$ $\frac{12}{14}$

Shade each model to show I. Name each fraction.

12 □/□

13 □/□

---

14   WRITING IN ▶MATH   Look at the model.
Name the fraction in two ways. Explain.

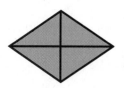

_____

_____

Vocabulary Check   Complete.

15   A _____ names part of a whole or part
of a set.

190   one hundred ninety

STOP

Name _____

Comparing Fractions

Key Concept

You can compare fractions using models.

$\frac{2}{3}$ $\frac{1}{3}$

$\frac{2}{3}$ of a pie is more than $\frac{1}{3}$ of a pie.

The fraction $\frac{2}{3}$ **is greater than** the fraction $\frac{1}{3}$.

Vocabulary

is greater than (>)   5 is greater than 1

  5 > 1

$\frac{3}{4}$ is greater than $\frac{1}{4}$.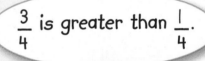

is less than (<)   2 is less than 10

2 < 10

The denominators are the same, so compare the numerators. 5 > 2

$\frac{5}{9} > \frac{2}{9}$ $\longleftarrow$ numerator $\longleftarrow$ denominator

Example

Compare the fractions. Write >, <, or =.

$\frac{2}{8}$ $<$ $\frac{5}{8}$

Step 1 Shade the first fraction. Shade 2 out of 8 parts.
Step 2 Shade the second fraction. Shade 5 out of 8 parts.
Step 3 Compare the numerators. 2 < 5

Answer $\frac{2}{8}$ $<$ $\frac{5}{8}$ $\frac{2}{8}$ is less than $\frac{5}{8}$.

Step-by-Step Practice

Compare the fractions. Write >, <, or =.

 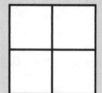

$\frac{3}{4}$ $\bigcirc$ $\frac{1}{4}$

Step 1 Shade the first fraction.

Shade _____ out of _____ parts.

Step 2 Shade the second fraction.

Shade _____ out of _____ parts.

Step 3 Compare the numerators. _____ $\bigcirc$ _____

Answer $\frac{3}{4}$ $\bigcirc$ $\frac{1}{4}$ $\frac{3}{4}$ is _____ than $\frac{1}{4}$.

Name _____

 Guided Practice

Shade and compare the fractions. Write >, <, or =.

1

$\frac{3}{6}$ ◯ $\frac{3}{6}$

2

$\frac{4}{10}$ ◯ $\frac{6}{10}$

Problem-Solving Practice

3 José jogged $\frac{1}{5}$ of a mile. Shawon jogged $\frac{1}{2}$ of a mile. Who jogged farther?

Understand   Underline key words.

Plan   Use fraction bars.

Solve

1   **1**

$\frac{1}{5}$ $\frac{1}{5}$

$\frac{1}{2}$  $\frac{1}{2}$

$\frac{1}{5}$ ◯ $\frac{1}{2}$

_____ jogged farther.

Check   Draw a picture. Which fraction has a greater area shaded?

▶ Practice on Your Own

Compare the fractions. Write >, <, or =.

4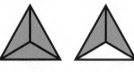

$\frac{3}{3}$ ◯ $\frac{2}{3}$

5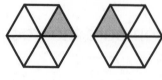

$\frac{1}{6}$ ◯ $\frac{1}{6}$

Shade and compare the fractions. Write >, <, or =.

6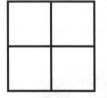

$\frac{2}{4}$ ◯ $\frac{3}{4}$

7

$\frac{3}{7}$ ◯ $\frac{1}{7}$

---

8 **WRITING IN ▶MATH** Benito says $\frac{1}{3}$ is greater than $\frac{1}{2}$. Is he correct? Explain.

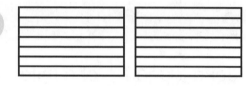

_____

_____

Vocabulary Check   Complete.

9 The symbol > means is _____ than.

10 The symbol < means is _____ than.

STOP

Name _____

Progress Check 1 (Lessons 6-1 and 6-2)

Name each fraction. Circle each fraction that equals 1.

1

2

3

$$\frac{\square}{\square}$$ $$\frac{\square}{\square}$$ $$\frac{\square}{\square}$$

Shade each model to show 1. Write the fraction.

4

5

6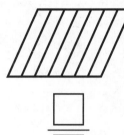

$$\frac{\square}{\square}$$ $$\frac{\square}{\square}$$ $$\frac{\square}{\square}$$

Shade and compare the fractions. Write >, <, or =.

7

$$\frac{3}{4} \bigcirc \frac{2}{4}$$

8

$$\frac{5}{6} \bigcirc \frac{3}{6}$$

9   Sam drank $\frac{1}{5}$ of a juice box.

Reid drank $\frac{3}{5}$ of a juice box.

Who drank more juice?

_____ drank more juice.

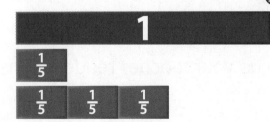

Name _____

Materials
2 game pieces
number cube labeled 1–6
blank number cube
(label the sides 8, 8, 10, 10, 12, 12)
eighth, tenth, and twelfth fraction strips

How to Play

Listen as your teacher reads the instructions.

Name _____

Equivalent Fractions

Copyright © Macmillan/McGraw-Hill • Glencoe, a division of The McGraw-Hill Companies, Inc.

Key Concept

Equivalent fractions represent an *equal* amount.

Use fraction bars to find equivalent fractions.

$\frac{1}{2} = \frac{2}{4}$

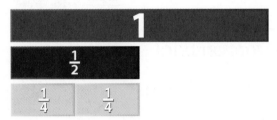

Vocabulary

equivalent fractions fractions that equal the same amount

$\frac{1}{2}$ = $\frac{2}{4}$

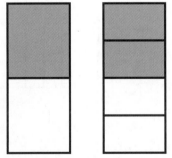

$\frac{1}{2}$ is the same as $\frac{2}{4}$.

The same area is shaded in $\frac{1}{2}$ and in $\frac{2}{4}$.

The fractions $\frac{1}{2}$ and $\frac{2}{4}$ are equivalent.

Example

Write an equivalent fraction.

Step 1 Shade the top fraction bar. Shade 1 out of 3 parts.

Step 2 Shade an equal amount of the bottom fraction bar.

Step 3 There are 2 parts shaded on the bottom fraction bar.

Step 4 Write the equivalent fraction.

$\frac{1}{3}$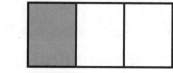

$\frac{2}{6}$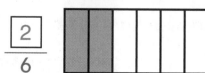

Answer $\frac{1}{3} = \frac{2}{6}$

Step-by-Step Practice

Write an equivalent fraction.

Step 1 Shade the top fraction bar. Shade __3__ out of __4__ parts.

Step 2 Shade an equal amount of the bottom fraction bar.

Step 3 There are _____ parts shaded on the bottom fraction bar.

Step 4 Write the equivalent fraction.

$\frac{3}{4}$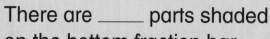

$\frac{\square}{8}$

Answer $\frac{3}{4} = \frac{\square}{8}$

Name _____

Write an equivalent fraction.

1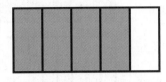

$$\frac{4}{5} = \frac{\Box}{10}$$

2

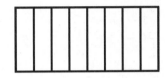

$$\frac{1}{2} = \frac{\Box}{8}$$

Problem-Solving Practice

3 Mercedes wants to color $\frac{1}{2}$ of the figure blue. How many parts should she color?

Understand   Underline key words.

Plan   Use fraction bars to find an equivalent fraction.

Solve   Mercedes should color

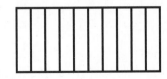

_____ parts blue. $\frac{1}{2} = \frac{\Box}{10}$

Check   Are the shaded areas equal?

GO on

Copyright © Macmillan/McGraw-Hill, • Glencoe, a division of The McGraw-Hill Companies, Inc.

Practice on Your Own

Write an equivalent fraction.

4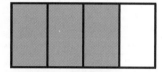

$$\frac{3}{4} = \frac{\square}{8}$$

5

$$\frac{1}{2} = \frac{\square}{6}$$

6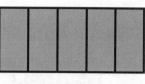

$$\frac{5}{5} = \frac{\square}{10}$$

7

$$\frac{2}{3} = \frac{\square}{12}$$

8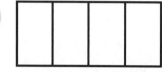

$$\frac{2}{4} = \frac{\square}{12}$$

9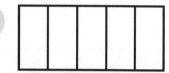

$$\frac{3}{5} = \frac{\square}{10}$$

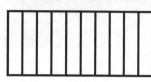

10 **WRITING IN ►MATH** Rita ran $\frac{4}{5}$ of a mile. Ed ran $\frac{8}{10}$ of a mile. Who ran farther? Explain.

_____

Vocabulary Check Complete.

11 Fractions that equal the same amount are

_____ .

STOP

Name _____

Fractions and Measurement

Key Concept

You can **measure length** to the nearest half-**inch**.

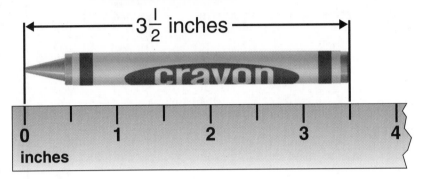

$3\frac{1}{2}$ inches

The length is halfway between the 3 and the 4.

The crayon is $3\frac{1}{2}$ inches long.

Vocabulary

measure   to find the length, height, or weight of an object

length   how long something is

inch   a customary unit for measuring length
The plural is *inches.*

length

The marks on a ruler are similar to a number line. Use the marks to find the length of an object.

Example

Measure to the nearest half-inch.

Step 1 Line up one end of the object with 0.

Step 2 Find the half-inch mark that is closest to the end of the object. $2\frac{1}{2}$ inches

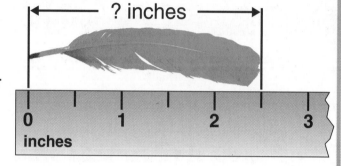

? inches

0 1 2 3
inches

Step 3 Write the length to the nearest half-inch.

$2\frac{1}{2}$ inches

Answer The feather is $2\frac{1}{2}$ inches long.

Step-by-Step Practice

Measure to the nearest half-inch.

Step 1 Line up one end of the object with 0.

Step 2 Find the half-inch mark that is closest to the end of the object.

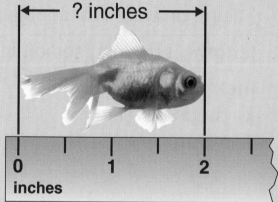

? inches

0 1 2
inches

_____ inches

Step 3 Write the length to the nearest half-inch.

_____ inches

Answer The fish is _____ inches long.

Name _____

 Guided Practice

Measure to the nearest half-inch.

1 _____ inches

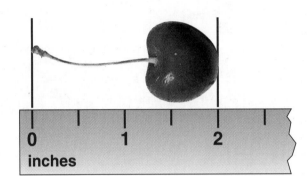

2 _____ inches

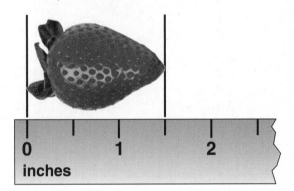

Problem-Solving Practice

3 Obike measures the carrot. What is the length to the nearest half-inch?

Understand Underline key words.

Plan Use a ruler.

Solve

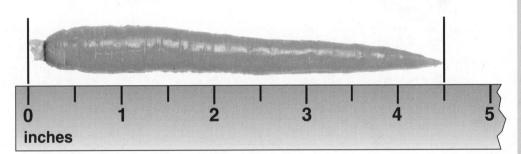

The carrot is _____ inches long.

Check Does one end of the carrot line up with 0?

GO on

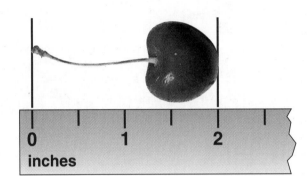

▶ Practice on Your Own

Measure to the nearest half-inch.

4 _____ inches

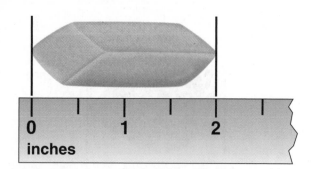

5 _____ inch

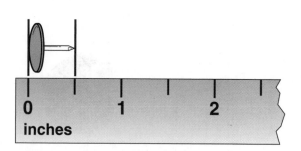

6 **WRITING IN ▶MATH**   Glen says the scissors are 4 inches long to the nearest half-inch. Is Glen correct? Explain.

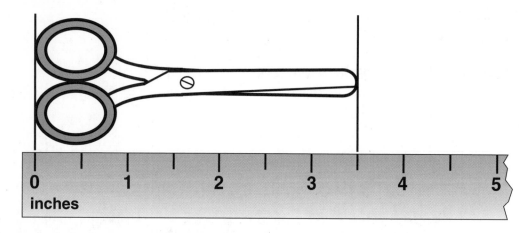

_____

_____

Vocabulary Check   Complete.

7 _____ is how long something is.

STOP

Name _____

Progress Check 2 (Lessons 6-3 and 6-4)

Write an equivalent fraction.

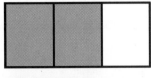

1

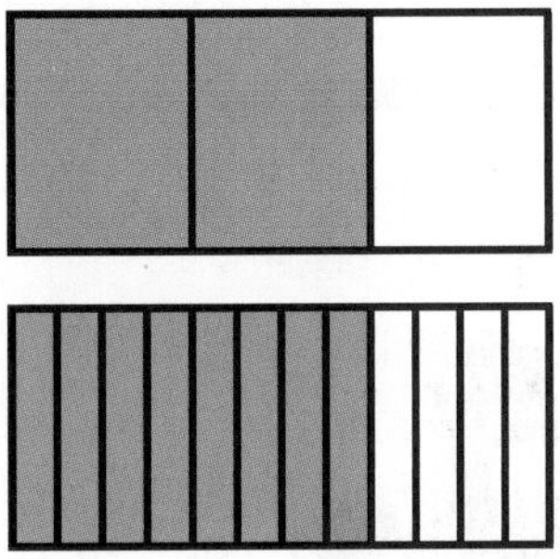

$\frac{2}{3} = \frac{\square}{12}$

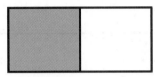

2

$\frac{1}{2} = \frac{\square}{8}$

3

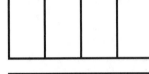

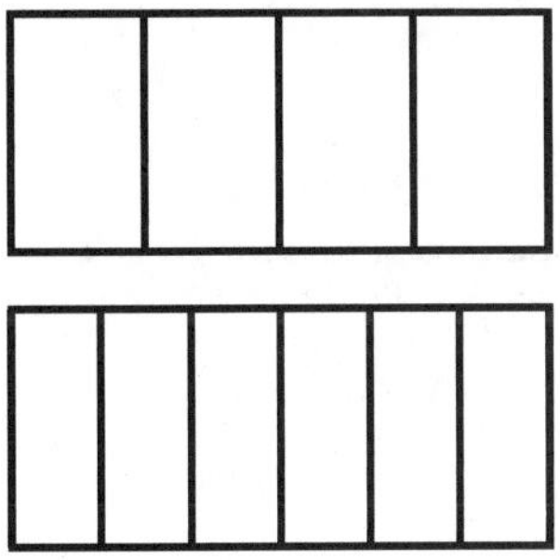

$\frac{2}{4} = \frac{\square}{6}$

4

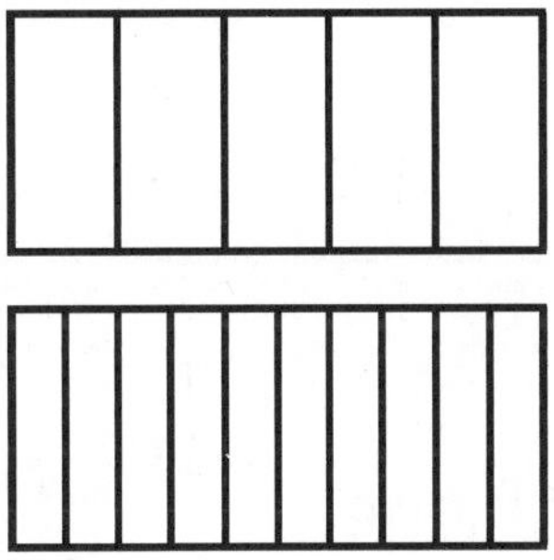

$\frac{4}{5} = \frac{\square}{10}$

---

5 Measure to the nearest half-inch.

_____ inches

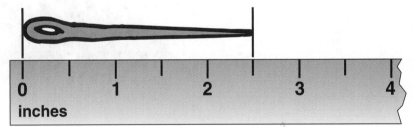

---

6 Adeola drew a green line. Use a ruler to measure the line to the nearest half-inch.

_____ inches

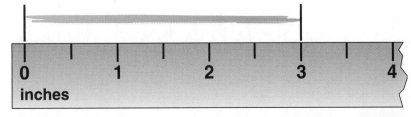

Name _____

«« Replay **What kind of worm can measure?**

Measure each object to the nearest half-inch.
Fill in the letters to answer the riddle.

_____ inches

_____ inches

_____ inches

_____ inches

An ☐ ☐ ☐ ☐ worm!

      $3\frac{1}{2}$ $1\frac{1}{2}$ 3 2

Name _____

Common Denominators

Key Concept

You can compare fractions with **common denominators.**

$\frac{2}{5}$ are quarters.

$\frac{4}{5}$ are quarters.

$$2 < 4$$

$$\text{So, } \frac{2}{5} < \frac{4}{5}.$$

Vocabulary

common denominator the same denominator in two fractions $\frac{2}{6}$ $\frac{3}{6}$

6 is greater than 2, so $\frac{6}{7} > \frac{2}{7}$.

numerator the top number in a fraction

denominator the bottom number in a fraction

$\frac{3}{5}$ ←numerator
←denominator

When the denominators are the same, the fractions are divided into the same number of equal parts.

Example

Compare. Write >, <, or =.

$\dfrac{3}{6}$ are blue.

$\dfrac{1}{6}$ is blue.

Step 1   Look at the denominators of the fractions. They are the same.

Step 2   Compare the numerators. 3 > 1

Step 3   Compare the fractions. $\dfrac{3}{6} > \dfrac{1}{6}$

Answer $\dfrac{3}{6}$ is greater than $\dfrac{1}{6}$.

Step-by-Step Practice

Compare. Write >, <, or =.

$\dfrac{3}{9}$ are green.

$\dfrac{7}{9}$ are green.

Step 1   Look at the denominators of the fractions. They are the same.

Step 2   Compare the numerators. _____ $\bigcirc$ _____

Step 3   Compare the fractions. $\dfrac{3}{9} \bigcirc \dfrac{7}{9}$

Answer $\dfrac{3}{9}$ is _____ than $\dfrac{7}{9}$.

Name _____

 Guided Practice

1. Compare. Write >, <, or =.

 $\dfrac{1}{4} \bigcirc \dfrac{3}{4}$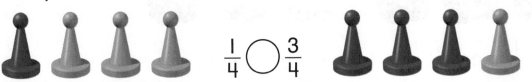

Problem-Solving Practice

2. Julia and Ann each have 10 state quarters. Julia's collection is $\dfrac{7}{10}$ Texas quarters. Ann's collection is $\dfrac{2}{10}$ Texas quarters. Who has fewer Texas quarters?

Understand   Underline key words.

Plan   Draw a model.

Solve   Shade _____ quarters in Julia's collection.

Shade _____ quarters in Ann's collection.

○○○○○   ○○○○○
○○○○○   ○○○○○

Julia's Quarters   Ann's Quarters

_____ has fewer Texas quarters.

Check   Compare the numerators.

▶ Practice on Your Own

Compare. Write >, <, or =.

3

$\frac{1}{3}$ is green.　　　　$\frac{2}{3}$ are green.

$\frac{1}{3}$ ◯ $\frac{2}{3}$

4 $\frac{8}{8}$ ◯ $\frac{8}{8}$　　　　**5** $\frac{2}{10}$ ◯ $\frac{9}{10}$　　　　**6** $\frac{4}{7}$ ◯ $\frac{6}{7}$

7 $\frac{6}{12}$ ◯ $\frac{11}{12}$　　　　**8** $\frac{6}{6}$ ◯ $\frac{4}{6}$　　　　**9** $\frac{3}{5}$ ◯ $\frac{1}{5}$

10 **WRITING IN** ▶**MATH** Paul ate $\frac{3}{6}$ of a pack of crackers. Kim ate $\frac{2}{6}$ of a pack of crackers. Who ate more crackers? Explain.

_____

Vocabulary Check Complete.

11 In the fraction $\frac{3}{5}$, 3 is the _____.

12 In the fraction $\frac{3}{5}$, 5 is the _____.

STOP

210 two hundred ten

Name _____

Common Numerators

Key Concept

Think about the size of the fraction parts to compare fractions.

 $\dfrac{1}{3}$

This circle is divided into 3 parts.

 $\dfrac{1}{8}$

This circle is divided into 8 parts.

The red piece is larger than the green piece.

$$\dfrac{1}{3} \; \bigcirc{>} \; \dfrac{1}{8}$$

Vocabulary

numerator the top number in a fraction

$\dfrac{3}{5}$ ←——numerator
←——denominator

denominator the bottom number in a fraction

is greater than (>) $\dfrac{1}{3}$ is greater than $\dfrac{1}{8}$ $\dfrac{1}{3} > \dfrac{1}{8}$

is less than (<) $\dfrac{3}{6}$ is less than $\dfrac{5}{6}$ $\dfrac{3}{6} < \dfrac{5}{6}$

One part of a whole that has only 3 parts is bigger than one part of a whole that has 8 parts.

When the numerators are the same, think about the size of the fraction parts.

Example

Shannon ran $\frac{1}{6}$ of a mile.

Robert ran $\frac{1}{2}$ of a mile.

Who ran farther?

 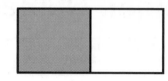

$\frac{1}{6}$ $\frac{1}{2}$

Step 1   Shade each fraction.

Step 2   Compare the shaded areas.

The $\frac{1}{6}$ part is less than the $\frac{1}{2}$ part.

Step 3   Write the correct symbol. $\frac{1}{6} \left(<\right) \frac{1}{2}$

Answer   Robert ran farther.

Step-by-Step Practice

Lucy painted for $\frac{1}{5}$ of an hour. Nick painted for $\frac{1}{10}$ of an hour. Who painted for less time?

Step 1   Shade each fraction.

Step 2   Compare the shaded areas.

The $\frac{1}{5}$ part is _____

than the $\frac{1}{10}$ part.

$\frac{1}{5}$ $\frac{1}{10}$

Step 3   Write the correct symbol.

$\frac{1}{5} \bigcirc \frac{1}{10}$

Answer _____ painted for less time.

Name _____

 Guided Practice

Shade the figures and compare. Write >, <, or =.

1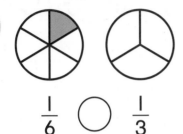

$\frac{1}{6}$ ◯ $\frac{1}{3}$

2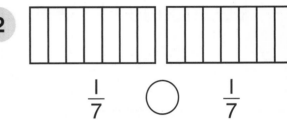

$\frac{1}{7}$ ◯ $\frac{1}{7}$

Problem-Solving Practice

3 Lara needs $\frac{1}{4}$ of a cup of flour and $\frac{1}{8}$ of a cup of sugar.

Does Lara need more flour or sugar?

Understand   Underline key words.

Plan   Use fraction bars to compare.

Solve

The _____ bar is larger than the _____ bar.

Lara needs more _____.

Check   Are the parts bigger in a fraction divided into 4 parts or 8 parts?

Practice on Your Own

Compare. Write >, <, or =.

4

$\dfrac{1}{10}$ ◯ $\dfrac{1}{4}$

5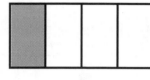

$\dfrac{1}{3}$ ◯ $\dfrac{1}{5}$

Shade the figures. Compare. Write <, > or =.

6

$\dfrac{1}{9}$ ◯ $\dfrac{1}{5}$

7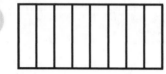

$\dfrac{1}{8}$ ◯ $\dfrac{1}{2}$

8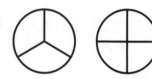

$\dfrac{1}{3}$ ◯ $\dfrac{1}{4}$

---

9 **WRITING IN** ▶**MATH** Gustavo says $\dfrac{1}{10} > \dfrac{1}{2}$.

Is Gustavo correct? How do you know?

_____

_____

_____

Vocabulary Check Complete.

10 The _____ tells how many equal parts a whole is divided into.

Copyright © Macmillan/McGraw-Hill • Glencoe, a division of The McGraw-Hill Companies, Inc.

(STOP)

Name _____

Progress Check 3 (Lessons 6-5 and 6-6)

Compare. Write >, <, or =.

1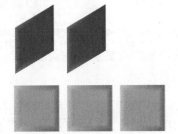

$\frac{2}{5}$ are blue.

$\frac{2}{5}$ ◯

$\frac{4}{5}$ are blue.

$\frac{4}{5}$

2 $\frac{2}{3}$ ◯ $\frac{1}{3}$ **3** $\frac{4}{6}$ ◯ $\frac{5}{6}$ **4** $\frac{7}{8}$ ◯ $\frac{2}{8}$

Shade the figures and compare. Write >, <, or =.

5

$\frac{1}{3}$ ◯ $\frac{1}{8}$

6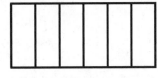

$\frac{1}{6}$ ◯ $\frac{1}{4}$

7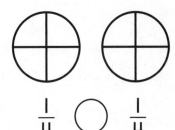

$\frac{1}{4}$ ◯ $\frac{1}{4}$

---

8 Kiah wants to make cinnamon bread using her grandma's recipe. Does Kiah need more water or milk?

Kiah needs more _____.

Grandma's Cinnamon Bread	
$\frac{1}{3}$ cup sugar	2 cups flour
$\frac{1}{3}$ cup water	2 eggs
$\frac{1}{2}$ cup milk	3 Tbs. butter
	4 Tbs. cinnamon

Name _____

A-maze-ing Treasure

Begin at START.
Compare each pair of fractions.
Follow the direction of the correct symbol.
End at the TREASURE!

| $\frac{1}{10}$ | $\frac{1}{10}$ | $\frac{1}{10}$ | $\frac{1}{10}$ | $\frac{1}{10}$ | $\frac{1}{10}$ | $\frac{1}{10}$ | $\frac{1}{10}$ | $\frac{1}{10}$ | $\frac{1}{10}$ |

| $\frac{1}{8}$ | $\frac{1}{8}$ | $\frac{1}{8}$ | $\frac{1}{8}$ | $\frac{1}{8}$ | $\frac{1}{8}$ | $\frac{1}{8}$ | $\frac{1}{8}$ |

| $\frac{1}{6}$ | $\frac{1}{6}$ | $\frac{1}{6}$ | $\frac{1}{6}$ | $\frac{1}{6}$ | $\frac{1}{6}$ |

| $\frac{1}{5}$ | $\frac{1}{5}$ | $\frac{1}{5}$ | $\frac{1}{5}$ | $\frac{1}{5}$ |

| $\frac{1}{4}$ | $\frac{1}{4}$ | $\frac{1}{4}$ | $\frac{1}{4}$ |

| $\frac{1}{3}$ | $\frac{1}{3}$ | $\frac{1}{3}$ |

| $\frac{1}{2}$ | $\frac{1}{2}$ |

| 1 |

START

Maze grid (each cell compares a pair of fractions):

Row 1:
- $\frac{3}{6} \bigcirc \frac{4}{6}$ < → / > ↓
- $\frac{1}{2} \bigcirc \frac{1}{3}$ < → / > ↓
- $\frac{2}{5} \bigcirc \frac{3}{5}$ < → / > ↓
- ← < $\frac{7}{10} \bigcirc \frac{3}{10}$ > ↓

Row 2:
- $\frac{1}{5} \bigcirc \frac{2}{5}$ < → / > ↓
- $\frac{1}{5} \bigcirc \frac{3}{5}$ < → / > ↓
- $\frac{1}{10} \bigcirc \frac{1}{5}$ < → / > ↓
- < ↑ $\frac{7}{8} \bigcirc \frac{3}{8}$ > ↓

Row 3:
- $\frac{1}{5} \bigcirc \frac{1}{8}$ < → / > ↓
- ← < $\frac{9}{10} \bigcirc \frac{6}{10}$ > ↓
- ← < $\frac{2}{4} \bigcirc \frac{3}{4}$ > ↓
- ← < $\frac{1}{6} \bigcirc \frac{1}{5}$ > ↓

Row 4:
- TREASURE!
- ← < $\frac{1}{8} \bigcirc \frac{1}{6}$ > →
- > ↑ $\frac{1}{6} \bigcirc \frac{1}{4}$ < →
- < ↑ > ← $\frac{1}{8} \bigcirc \frac{1}{10}$

216 two hundred sixteen

Name _____

Review

Vocabulary

Word Bank

denominator

equivalent
fractions

numerator

<

>

Use the Word Bank to complete.

1 $\frac{3}{5}$ ◄·········· _____

2 $\frac{1}{3} = \frac{2}{6}$ _____

3 $\frac{1}{4} \bigcirc \frac{3}{4}$

4 $\frac{1}{2} \bigcirc \frac{1}{3}$

5 $\frac{6}{8}$ ◄·········· _____

▶ Concepts

Name each fraction. Circle each fraction that equals 1.

6

$\frac{5}{5}$

7

$\frac{\square}{\square}$

8

$\frac{\square}{\square}$

Shade each model to show 1. Name each fraction.

9   $\frac{\square}{\square}$

10   $\frac{\square}{\square}$

GO on

Copyright © Macmillan/McGraw-Hill, • Glencoe, a division of The McGraw-Hill Companies, Inc.

Shade and compare the fractions. Write >, <, or =.

11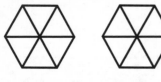

$$\frac{1}{6} \bigcirc \frac{5}{6}$$

12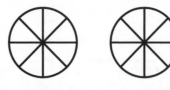

$$\frac{2}{8} \bigcirc \frac{2}{8}$$

Write an equivalent fraction.

13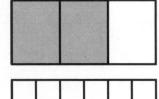

$$\frac{2}{3} = \frac{\square}{6}$$

14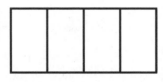

$$\frac{2}{4} = \frac{\square}{\square}$$

15 Measure to the nearest half-inch.

_____ inches

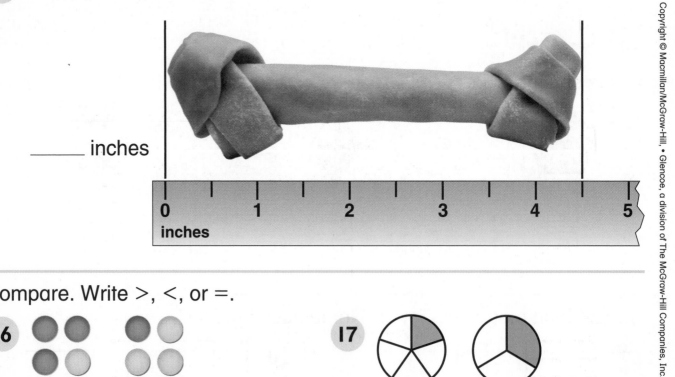

Compare. Write >, <, or =.

16

$$\frac{3}{4} \bigcirc \frac{1}{4}$$

17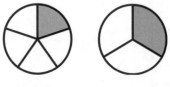

$$\frac{1}{5} \bigcirc \frac{1}{3}$$

218 two hundred eighteen

Review

STOP

Name _____

Chapter Test

Compare. Write >, <, or =.

1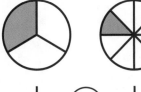

$\frac{3}{5}$ ◯ $\frac{2}{5}$

2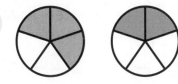

$\frac{3}{8}$ ◯ $\frac{6}{8}$

3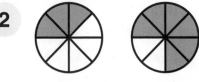

$\frac{1}{3}$ ◯ $\frac{1}{8}$

4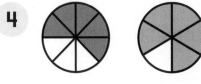

$\frac{5}{8}$ ◯ $\frac{5}{6}$

Write an equivalent fraction.

5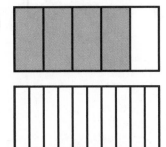

$\frac{4}{5} = \frac{\square}{10}$

6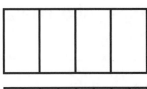

$\frac{3}{4} = \frac{\square}{\square}$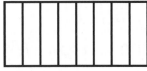

---

7 Measure to the nearest half-inch.

_____ inches

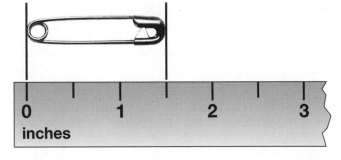

8 Who is Correct?

Walter and Cindy find fractions greater than $\frac{1}{5}$.

$\frac{1}{10}$ is greater than $\frac{1}{5}$.

Walter

$\frac{3}{5}$ is greater than $\frac{1}{5}$.

Cindy

Circle the correct answer. Explain.

_____

_____

---

9 Sandra rode her bike for $\frac{4}{5}$ of an hour. Jon rode his bike for $\frac{2}{5}$ of an hour. Who rode for a longer time? _____ rode for a longer time.

4 > 2, so $\frac{4}{5}$ > $\frac{2}{5}$.

10 Look at the window. What fraction represents the whole window? _____

11 Jasmine walked $\frac{1}{4}$ of a mile. Henry walked $\frac{1}{6}$ of a mile. Who walked less distance? Explain.

_____

_____

STOP

Name _____

Test Practice

Choose the correct answer.

1   Molly's mother cut an apple into 3 equal pieces. Molly ate the whole apple. What fraction of the apple did Molly eat?

   ○ $\dfrac{3}{1}$ ○ $\dfrac{3}{3}$ ○ $\dfrac{1}{3}$ ○ $\dfrac{1}{2}$

2   How long is the candle to the nearest half-inch?

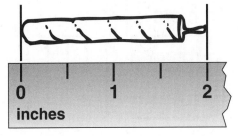

   ○ $\dfrac{1}{2}$ inch   ○ 1 inch

   ○ $1\dfrac{1}{2}$ inches   ○ 2 inches

3   Complete.

   $\dfrac{2}{6}$ ○ $\dfrac{2}{3}$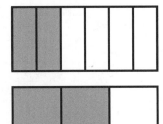

   ○ <   ○ +   ○ =   ○ >

4   Paul baked banana bread. He sliced the bread into 6 pieces. What fraction represents the whole loaf of bread?

   ○ $\dfrac{6}{6}$ ○ $\dfrac{6}{1}$ ○ $\dfrac{1}{6}$ ○ 6

5   Which number sentence is true?

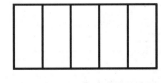

   ○ $\dfrac{1}{5} = \dfrac{1}{4}$ ○ $\dfrac{1}{4} > \dfrac{1}{5}$

   ○ $\dfrac{1}{4} < \dfrac{1}{5}$ ○ $\dfrac{1}{5} > \dfrac{1}{4}$

6   Which fraction equals 1?

   ○ $\dfrac{1}{5}$ ○ $\dfrac{2}{5}$

   ○ $\dfrac{4}{5}$ ○ $\dfrac{5}{5}$

7 Which number sentence is true?

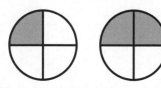

○ $\frac{2}{4} < \frac{1}{4}$ ○ $\frac{2}{4} = \frac{1}{4}$

○ $\frac{1}{4} > \frac{2}{4}$ ○ $\frac{2}{4} > \frac{1}{4}$

---

8 How long is the key to the nearest half-inch?

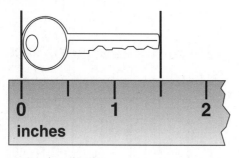

○ $\frac{1}{2}$ inch   ○ 1 inch

○ 1 $\frac{1}{2}$ inches   ○ 2 inches

---

9 Carlos gave away $\frac{3}{5}$ of his model cars. Which fraction is equivalent to $\frac{3}{5}$?

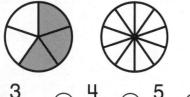

$\frac{3}{10}$ ○ $\frac{4}{10}$ ○ $\frac{5}{10}$ ○ $\frac{6}{10}$

---

10 Which number sentence is true?

○ $\frac{5}{8} = \frac{3}{8}$ ○ $\frac{3}{8} > \frac{5}{8}$

○ $\frac{3}{8} < \frac{5}{8}$ ○ $\frac{5}{8} < \frac{3}{8}$

---

11 Holly ate $\frac{1}{2}$ of her sandwich. Tom ate the same amount of his sandwich. His sandwich was cut into 8 pieces. What fraction of his sandwich did Tom eat?

○ $\frac{2}{8}$ ○ $\frac{4}{8}$ ○ $\frac{3}{8}$ ○ $\frac{5}{8}$

---

12 Which number sentence is true?

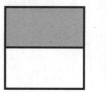

○ $\frac{1}{4} > \frac{4}{10}$ ○ $\frac{4}{10} > \frac{1}{4}$

○ $\frac{4}{10} = \frac{1}{4}$ ○ $\frac{4}{10} < \frac{1}{4}$

STOP